ther Pergamon Titles of Interest

ILLINGTON & ROBERTS	Building Services Engineering: A Review of its Development
ANCIS	Introducing Structures
BSON	Thin Shells: Computing and Theory
RRISON	Structural Analysis and Design: Some Minicomputer Applications
BS & DOLING	Planning for Engineers and Surveyors
NE	Plastic Theory of Structures, 2nd Edition
R	Solar Energy Applications in Houses
AWAY	Engineers in Industry
ZNER	Movements in Buildings, 2nd Edition
S *et al*	Aspects of Civil Engineering Contract Procedure, 2nd Edition
AGHAN	Building for Energy Conservation
AGHAN	Design and Management for Energy Conservation
STEEMERS	Solar Houses in Europe: How They Have Worked
T	Initial Skills in Bricklaying
	Energy Resources and Conservation in the Built Environment (2 volumes)

Related Journals - *free specimen copies sent on request*

Environment
oncrete Research
Structures
ion and Management
tional
rnal of Solids and Structures
Recovery Systems

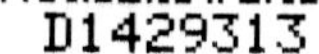

PERGAMON INTERNATIONAL LIBRAR
of Science, Technology, Engineering and Socia

The 1000-volume original paperback library in aid o
industrial training and the enjoyment of le

Publisher: Robert Maxwell, M.C.

BUILDING ECON

Third Edition

THE
INS

An inspection copy of a
sent to academic sta
recommendation. Co
not suitable. When a
the recommendatio
with our complime
and new titles to

B

FR
GI
HA

HO
HO
JAGE
KENN
LENC
MARK

O'CAL
O'CALL
PALZ &
TEMPES
URAL

Pergamon

Building and
Cement and C
Computers and
Energy Convers
Habitat Interna
International Jo
Journal of Heat
Solar Energy

BUILDING ECONOMY

Design, Production and Organisation

A Synoptic View

THIRD EDITION

by

P. A. STONE M.Sc. (Econ.), Ph.D.

PERGAMON PRESS

OXFORD · NEW YORK · TORONTO · SYDNEY · PARIS · FRANKFURT

U.K.	Pergamon Press Ltd., Headington Hill Hall, Oxford OX3 0BW, England
U.S.A.	Pergamon Press Inc., Maxwell House, Fairview Park, Elmsford, New York 10523, U.S.A.
CANADA	Pergamon Press Canada Ltd., Suite 104, 150 Consumers Rd., Willowdale, Ontario M2J 1P9, Canada
AUSTRALIA	Pergamon Press (Aust.) Pty. Ltd., P.O. Box 544, Potts Point, N.S.W. 2011, Australia
FRANCE	Pergamon Press SARL, 24 rue des Ecoles, 75240 Paris, Cedex 05, France
FEDERAL REPUBLIC OF GERMANY	Pergamon Press GmbH, 6242 Kronberg-Taunus, Hammerweg 6, Federal Republic of Germany

First edition 1966
Second edition 1976
Third edition 1983

British Library Cataloguing in Publication Data

Stone, P. A.
Building economy.—3rd ed.—(Pergamon International Library)
1. Construction industry
I. Title
338.4′ 769 HD9715.A2

ISBN 0-08-028677-1 Hardcover
ISBN 0-08-028678-X Flexicover

Printed in Great Britain by A. Wheaton & Co. Ltd., Exeter

To My Wife

CONTENTS

PREFACE TO THE THIRD EDITION

THE changes which have taken place in the construction industry, since the second edition of this book was prepared, have continued the previous evolutionary trends. Major changes have occured outside the industry; in the world and local economy with effects on demand and prices, in the development of new technologies such as electronics with affects on demand and aids available to the industry, and in population size, structure and movements, and styles of living with effects on the nature and scales of demand. Such changes affect the environment in which the industry operates rather than the principles of design, construction and management with which this book is mainly concerned. These changes are, however, important for the structure of the industry and for strategies it might follow. The third edition is therefore being extended with an additional part, Part V1, with chapters dealing respectively with approaches to assessing future trends in the context for construction and future demand, and their possible consequences; and with a discussion of strategies which the industry and government might follow. An additional chapter is also provided illustrating the way the main principles of economics relate to the design, construction and management of building and buildings. The data and illustrations have been updated as far as possible.

1982 P.A.S.

PREFACE TO THE SECOND EDITION

DESPITE the changes which have taken place in the economy and in construction, in principle the arguments advanced in the First Edition of this book remain true. The problems of determining the most efficient design, method of production and organisation of the work of construction remain much the same. The major changes which have occurred have been to the economy, inflation, the phenomenal rise in energy prices and the decline in domestic output. The conditions vary from country to country and will vary over time. The change in economic climate does not change the principles. However, more attention is now being given to resource and especially energy problems and government influence, both of which have chapters to themselves. There is also an additional chapter reviewing the use of energy in and in connection with building and the economies which can be made in the use of energy and its relation to design and construction. All the original material has been revised in the light of change and additional material added as necessary.

1975 P.A.S.

PREFACE TO THE FIRST EDITION

IN THE long run buildings are judged in terms of the trinity of appearance, convenience, and cost. The economic building is not necessarily the cheapest building but the one that provides the best value for money, that is, the one that is of good appearance and convenient in relation to the costs of constructing, running and operating in it.

This book is concerned with building economy, with those aspects of design and production and the related problems of organisation which affect the costs of a building and which hence affect the money side of the value for money equation. It is, therefore, concerned with the cost aspects of design, construction and organisation and not with the aesthetic and technical aspects of these matters. The purpose of the book is to describe and assess the forms of construction used today, the methods of production and the way the industry is organised, and to consider the impact which new methods, materials and forms of organisation and contractural relationships might have on the industry. Thus it will be necessary to consider the importance of prefabrication, the use of system building and other forms of industrialisation, mechanisation, programming and other management techniques, and the alternative ways of relating the designers, manufacturers and constructors.

The methods of construction and the way building is organised differ from one country to another. Solutions economic in one country are not necessarily so in other countries. Some consideration will be given to the extent to which solutions valid in Britain can usefully be adopted in other countries.

The aim of the book is to provide an interpretation of the building economy which will aid the different disciplines in the fields of design and production, and building clients to appreciate the field as a whole and to relate the various parts of it. The book is written mainly as an introduction to professional students in the fields of building design, production, organisation and administration but may also interest professional people and general readers who require an up-to-date picture of the building economy.

I should like to thank Sir Fredrick M. Lea, C.B., C.B.E., the Director of Building Research, who kindly gave his approval to the preparation of this book when I was a member of his staff. I should also like to thank M. C. Fleming, B. A., the Research Fellow in Building Economics at the Queen's University, Belfast, and A. Gordon, Dip.Arch., F.R.I.B.A., who were kind enough to read the book in draft and who made some most valuable suggestions. Neither gentleman is, of course, in any way responsible for any shortcomings in the final result. I am also grateful to G. D. N. Worswick, M.A. Oxon., the Director of the National Institute of Economic and Social Research, who kindly read the book in proof.

1966 P.A.S.

CHAPTER ONE

INTRODUCTION

I

Building is one of the most important activities in any economy. A large part of the national resources are usually used in the construction and maintenance of buildings and buildings play an important part both in production and in providing services to the community. Clearly, it is important that good value should be obtained from the resources used. It is not suprising that building and buildings should be subject to frequent criticism. Unfortunately much of this criticism is ill-informed. All too frequently the complex of factors relating to building is ignored and as a result many of the recommendations are of little value or even damaging.

While many products, and the industries which produce them, have come to fruition in half a century, building and buildings as they are known today have been developing throughout recorded history. Their development has taken place both in time and geographically. The conical hut of branches is both the origin of contemporary building forms and in some parts of the world a contemporary building form itself. While the ordinary citizen built in mud, wattle and timber, the church and state built in brick and stone. Naturally, it is only examples of the monumental building of the church and state which survive today. However, while the buildings of the ordinary citizen have not endured, the history of the buildings is largely discernible in the contemporary forms of the less developed peoples of the world. A study of the relationships between the forms of construction and the contemporary conditions, both in the past and currently in other countries, is of assistance in obtaining an understanding of the current problems of the building economy.

Where the economy is poor the ordinary citizen builds for himself and professional builders are employed only by the church and state. Thus the two kinds of buildings, domestic and monumental, tend to develop in different ways. Domestic buildings tend to be built by the house-holders and their neighbours from such materials as lie ready to hand on the site. At first, materials are used in their crude state, but gradually more effort is made to fashion the materials to fit together to provide shelters of a more convenient form and better able to provide protection from the elements. Naturally, people who only occasionally take part in an activity cannot acquire the skill and knowledge of people practising full time. The skill of the professional is passed on from one generation to the next and is cross-fertilised by contact with builders from other areas. Professional building tends to advance much more rapidly than the building of the householder. As society becomes more wealthy and sophisticated, the proportion of people satisfied with shelters they can build themselves and with the need and time to erect and maintain their own shelters declines and more and more people employ professional builders. Thus the evolution of building and buildings is a result of the interaction between materials, skills and the external economies.

In many ways professional buildings have not changed radically over the last 2000 years. The houses of the wealthy Romans were not dissimilar to houses of today. For instance, both are based on similar materials, brick and timber, the sizes of the spaces in both are restricted by the span provided by the available timbers and often the forms of heating are not dissimilar. The engineering structures of the Romans were as large and as ambitious as those of today. On the other hand, the range of materials available today is far wider and the knowledge of their properties is greater. As a result the materials can be used more adventurously and economically. Moreover, the range of mechanical aids is far greater and so less labour and time needs to be used. The most significant changes in buildings lie in the range of engineering services now incorporated in them.

While in the more developed countries professional building accounts for most of the building work, the reverse is true in the less developed countries. In such countries homes and other small buildings are generally built by the owner and his friends from local materials, while professional building is mostly confined to commercial, communal and industrial buildings and works. Naturally, the owner-built dwellings usually lack the amenities expected in western type homes, especially the mechanical engineering and plumbing services. One of the economic problems in developing countries is to determine the best point at which to change from the use of self-built constructions, based on local and usually short life materials, to professional building based on manufactured materials.

While in sophisticated societies the ordinary citizen usually looks to the professional builder for major construction work, he frequently undertakes minor works and maintenance himself. This trend tends to grow in societies in which the cost of labour is growing rapidly relative to the cost of materials and components, and net after tax earnings, and in which changes in components and a growing availability of cheap small power tools reduces both the skill content and arduousness of the tasks.

II

The evaluation of buildings depends on whether they are required as a means of production or for consumption. In the productive sector of the economy a building is just another factor of production; it has no value in itself but only in as far as it provides shelter for a productive or distributive process or generates an income. It is a cost of production and has a precise money value. Consumers value buildings for themselves; for the satisfactions which they provide. Many of these satisfactions are personal and subjective, and no precise objective valuation is possible.

Buildings and works may be supplied either "bespoke", or "off-the peg"; the former are designed especially to meet the expressed needs of a particular client, while the latter are available either as speculatively erected buildings for sale, or as standard units for erection on the clients' sites. Most public, commercial and industrial buildings are of the first class, while most private housing is of the second class. However, increasingly standard units of buildings are being manufactured for housing, industrial and storage purposes and to some extent for educational, social and office purposes. On the other hand, civil engineering works are usually designed to the requirements of an individual client.

Bespoke buildings are mostly designed and erected by two separate groups of people, the materials being supplied by a further group. The design group, which may consist of several separate groups, usually under the co-ordination of an architect, are professionals, who act for the client on a fee basis. The contracting group, which may consist of a dozen or more separate firms, work for profit and generally secure the contracts on competitive basis. Again, the suppliers of the materials and components consist of many separate firms competing for the business. Sometimes they compete with designs as well as with the supply of the components. There is less division of responsibility in the case of off-the-peg buildings. The designer is employed by the contractor; often he is on his staff. Frequently, especially where the buildings are supplied in the form of standard units, the contractor also manufactures the major material or component. Thus in such cases the designer-producer relationship is similar to that in manufacturing industry, where the manufacturer is responsible for both design and production.

However, whichever type of building is used the contractor is largely an erector, an assembler of the products of other industries. Except for buildings based on unprocessed materials, about half the cost of a building is the cost of the materials used in its construction. This proportion rises as less use is made of the products of the extractive industries and more use is made of large prefabricated components.

III

The efficiency of the building industry and hence the cost of building depends on five main factors: the clients, the designers, the contractors, the producers of materials, and the economic and institutional environment. The client's contribution lies in his skill in specifying his needs prior to the preparation of the design and in assessing the value of the solutions put forward by the designer. The designer occupies the central role. His contribution lies in meeting the needs of the client with a solution economic to construct and to operate. He must thus take account of the problems likely to arise in the erection of the building and of the extent to which materials and components are or can be made available. Designers can have a larger influence on cost than either contractors or producers of materials since they determine the overall building and can reduce cost not only by using materials to the best advantage and in a way which simplifies erection but also by planning to meet the needs in a smaller space and by enclosing it in the minimum area of walls and roof. The contractors' contribution to cost reduction lies in the efficiency with which they assemble the building. This is not just a question of speed but of a proper balance between labour and organisation. The producers of materials can assist in cheapening the product by improving the efficiency of their process of extracting, manufacturing and distributing the materials and components. The economic and institutional environment affects efficiency by the way in which it facilitates the design and construction process and by the restrictions it imposes upon the industry. For instance, a good system of transport, an equitable climate and a well educated labour force all tend to assist the building industry to be efficient, while import and labour mobility restrictions and outdated safety and health regulations tend to reduce efficiency.

Most clients are involved in the exercise of their function only occasionally and must rely heavily on the skill of the designer in assessing the nature of their needs and the value of the solutions put forward.

The design of a building, particularly today, is a very complex problem, demanding a greater breadth of knowledge than most other design problems. The complexity arises from the size of a building and the number of purposes it serves, from the number and range of services to be provided and from the range of materials and methods of construction which are possible. Design economy can only be achieved through an understanding of the fundamental user requirements and an ability to compare the ultimate cost consequences of the many possible solutions. The design of different components of a building interact with each other and with the ways the building can be used, and affect the costs of running it and of operating within it. Few people can be experts in all the facets of the design of a building and its services but the leader of the design team must at least have an overall appreciation of the problems of all the disciplines.

The development of new materials and new building technqiues and the increase in the complexity of the services provided in a building has added to the erection problems as well as to the design problems. Not so long ago the range of materials and techniques of construction were so limited that building craftsmen could set out and erect buildings from the information given in simple outline drawings. The services to be fitted were few and simple and a contractor could generally understand the whole building process. Today the possible forms of construction are extensive and many of the services are very complex. Detailed drawings and often specialised knowledge are necessary. The traditional crafts tend to play a role of reduced importance and more and more specialist workers are employed. The more complicated forms of building and the demand for more rapid erection necessitate careful site planning and organisation; it is no longer adequate to follow through the natural sequence of trades. Moreover, the improvements in the standards of living and of general working conditions, and the tendency to labour shortages in many of the more advanced countries, have resulted in a need for better organisation and labour conditions and for higher pay. The greater the relative cost of site labour the greater the need for labour to be used efficiently.

Traditional building materials are generally bulky and heavy, and variable in shape and size. While cheap in themselves they are often expensive to use. Techniques of using them have been evolved which eliminate most of the difficulties arising from their lack of uniformity. Even so, such materials often have a limited technical performance. Moreover, being bulky and heavy they are costly to transport. At a time when transport was relatively undeveloped, and in countries where this is still so, building materials are not generally much used outside their locality of origin. Where such conditions pertain the forms of construction are limited and building is usually expensive. The development of cheap transport results in a cheapening of materials and increases the range available. It both encourages the spread and standardisation of existing materials and favours the development of easily transportable materials.

New materials, often developed in other fields, have been found which provide greater flexibility in use and which are lighter and more uniform than traditional materials. Such materials often enable a greater range of problems to be solved and are often cheaper for their purpose than traditional materials. Frequently, however, the newer materials do not combine easily with the traditional ones.

IV

Innovation in building has been generated in many different ways. Clients have set new problems; they have demanded better standards of comfort and convenience, better services and more economic solutions. Designers have looked to the potential of new materials to solve new problems and to solve old problems in new and better ways. Constructors have used the new materials in an effort to reduce the costs of construction. Their freedom to choose materials is only complete where they are building directly for the market. Building offers a large potential market and the producers of materials have naturally tried to exploit this market for their materials.

While innovations have added to the range of materials available for building, increased the number of possible techniques, and made it possible to provide a better standard of comfort and service, they have done little to reduce the costs of construction. Increases in the standards of living have tended to lead to a rise both in the national costs of building construction and in the costs of individual buildings. While rises in the efficiency in other sectors of industry may leave the potential purchasers of buildings with more resources to devote to this purpose, it is unlikely that demand can be met, unless higher efficiency and hence lower relative costs can be achieved.

Many observers have felt that traditional building, involving as it does the joining together of a large number of small units, must be basically inefficient. Building has been critically compared with the factory industries and with the success they have achieved in mechanisation, in replacing craft processes with machines and with semi-skilled labour and with their success with large scale methods. There have been various attempts to try to emulate these methods in building. Broadly, these attempts have taken four forms; prefabrication, system building, mechanisation, and the rationalisation of the erection process.

Prefabrication has developed along two lines: in the form of standard as compared with purpose-made components and in the form of systems of construction based on large scale purpose-made components. The use of standard components has been developing steadily over the last half century. More and more items of joinery and metal goods are purchased ready made; plasterboard and plaster panels are replacing wet plaster; electrical and plumbing systems can be obtained with units cut to size and ready for installation. The development of systems of construction has naturally been most noticeable in those fields in which a standardised product is acceptable, for example, housing and schools. Some of these systems have been based on interlocking units which form a load-bearing structure; the units vary in size from traditionally sized building blocks to room-sized units. Other systems have been built around a frame hung with a light cladding material.

Mechanisation has mainly developed in four directions: earth movement, materials handling, concrete mixing and powered hand tools to assist manual tasks. The greatest success has perhaps been obtained in earth moving, where the power and endurance of the machine in relation to its cost is so much greater than that of the man it replaces. Materials handling has developed in step with the growth in size of building units. Machine concrete mixing also replaces labour on a very arduous task. No real success has so far been achieved in mechanising craft processes, but the development of powered hand tools, jigs and simply installed prefabricated components has reduced both the skill content and the physical difficulties of much craft work. This tends to reduce the intensity of skill training necessary for craftsmen and increases the feasibility of training craftsmen able to handle a much wider range of tasks, reducing the number of separate operatives necessary for carrying out building operations.

The development of prefabricated methods of building and of mechanisation has necessarily been accompanied by the rationalisation of the erection process. This starts with the study of the flow of work on the site. Up to a point a building job can be preplanned so that all the operations can be timed to fit into one continuous sequence of operations. To be successful such planning must be accompanied by very good organisation so that all the various types of labour and materials are brought on to the site at the right time. The full potential of such methods cannot be realised unless the design and building methods are examined together. The substitution of certain materials or certain design features may greatly simplify the whole construction process, especially where complete operations can be eliminated. Thus successful rationalisation in the building industry requires an examination of the design and construction together and their development as an organic whole.

V

In the long run the design and production, together with the contractual procedures which link together the various parts of the industry, must be considered as a single organism, each part of which not only contributes to the final product but which goes towards determining the forms of product which best meet the requirements.

VI

Buildings and engineering works generally have a long life and considerable consequences for the locality in which they are sited. Broadly the costs of maintaining and servicing a building over its life are as great as the original costs of its development. The consequences arising from the nature, design and location of a building and for the processes carried on within it, can result in costs as significant as those for its development. Account of the probable pattern of use and demand for a building over its future life needs to be taken into account in its design and location. This involves the use of techniques such as cost planning and costs-in-use.

VII

The materials and components available to the construction industry change as existing supplies of natural resources are exhausted or become more costly to obtain and as innovation adds to the range available, and changes the relative economics of different options. Minerals have been a major source of building materials. As easily accessable supplies become exhausted and the supply and cost of energy needed to transport and process them increases their relative prices tend to rise. While new methods of processing and using mineral resources add to the supplies of suitable building materials, minerals cannot be regenerated. More attention may need to be given to use of vegetable materials, since these can be regenerated.

Not only is a large volume of energy used in preparing construction materials and in construction processes but even more is used in maintaining acceptable environments within buildings. As energy becomes more scarce and relatively expensive, more attention is likely to be given to designing buildings and equipment which are more economic in their use of energy.

VIII

The context within which construction arises and the demand for construction work changes with changes in such phenomena as population levels, structures and locations, life styles, standards, technology, and national and world economy. In the western world population tends to be stabilising but its age structure tends to be changing. There is a movement of population out of the large cities and metropolitan areas, and the countryside to smaller towns. Such movements tend to increase the volume of redundant buildings and works while creating an increased demand for new development in the areas to which the migrants go. Much of the stock of buildings and works is very old, particularly in the locations from which population is tending to move, and much of it has not been adequately maintained or adapted to future needs. A mounting backlog of work for replacement of old buildings and infrastructure appears to be building up. Changes in life styles and expectations of standards may also give rise to extensive new building and rehabilitation work. Technological change, now faster than ever, also creates a need for changed patterns of building types, forms and locations. At the same time a slow down in the world economy and financial and fiscal measures intended to deal with inflation and an excess of labour over demand, tends to reduce the resources available to finance construction and the demand for it.

Changes in the nature and level of demand for construction work will have consequences for the construction industry, for its scale, form and structure. If resources are not to be wasted, the industry needs a lengthy period in which to adapt to meet likely future demands on it. A study of the factors likely to influence future demand and its timing would provide a basis for projecting the future demand for each type of work and the requirements these would create for the various types of materials and components, design, operative and organisational skills, plant and finance.

Generally the public sector is both a major user of construction resources and a major determinant of the context for private demand and the environment in which the industry works. There is a need to review public policies in the light of likely future needs for construction work. Consideration is needed of such matters as the management of public orders, finance, taxation, planning, land and training.

PART I

The Economic Background

CHAPTER TWO

ECONOMIC PRINCIPLES AND THEIR APPLICATION TO CONSTRUCTION

THE science of economics is based on a consideration of rational reactions to the expenditure of scarce resources to obtain a maximum return. For the consumer the return is measured in terms of satisfactions for the time, energy or money spent. For the business man the return is money surplus or profit left after all expenses have been set against the sale price, or is expressed in terms of the return on the capital expended. Studies of different types of actions have resulted in general laws or principles as a guide to decision. These can often be expressed mathematically and given appropriate statistical observations, they can be expressed arithmetically. They can then be applied to similar situations to estimate the consequences of possible decisions. They thus relate to decision theory, operational method and other business decision techniques. The more perfect the knowledge of a situation and the deeper the analysis of possible operations, the more likely that the solution giving the best return will be found.

Economic theory is a large subject and it is only possible in this chapter to provide a brief introduction to some of the principles more likely to arise in handling building problems. Even a limited sample will indicate the nature of economic analysis and the way it can be applied to design, construction and production problems. It will be appreciated that economic analysis lies behind the conclusions presented in later chapters, although there the discussion is presented more in technological and managerial terms than in terms of economic analysis. Readers requiring a more complete account of basic economic analysis should read one of the economic textbooks listed (1 & 2) or a book relating economic theory to the construction industry.(3)

INCREASING RETURNS

Up to a point the addition of a factor of production can result in an increase in returns more than proportional to the increase in the factor. This can occur for a number of reasons such as an ability to make a better use of available skills, a better use of an indivisable factor of production and to substitute better skills or capital equipment. Some examples will clarify the principle. An operative working single handed to build a brick wall will need to carry out several separate operations such as assembling and placing materials, mixing mortar, laying bricks and measuring. The skill and rhythm of each operation will be reduced by frequent breaks to carry out the other operations. If several men are employed each can specialise in a particular skill and tool, and achieve greater productivity. At the same time less time will be spent on non-productive work. If the job is large enough together they should turn out more work than if each were working singly.

Again two or three men often cannot utilise the full potential of even a small concrete mixer or crane. If more people can be used on the job, potential can be used more fully and costs per unit output generally reduced.

The output of most types of plant increases more than in proportion to its costs as its size is increased; the economics of scale. Costs per unit can therefore be reduced generally by using a larger size of plant but only if the full potential can be used over a long period.

Again a larger scale of operation may justify use of a higher type of skill or more sophisticated type of plant reducing the unit costs of employing operative skills.

DECREASING RETURNS

Increasing the scale of operation beyond a certain point may result in decreasing returns. This could occur, either if so many units of production were applied to a fixed unit that utilisation per unit was reduced or if overhead costs were pushed up. While a larger gang could make a better use of the output of a concrete mixer and utilise more of its potential, if there were too many in the gang they would tend to impede each other, perhaps in moving concrete from the mixer to the job or in pouring the concrete itself. Clearly the job needs to be large enough to enable a larger scale of operation to be economic. It is not, however, sufficient to limit consideration to the operation itself. As will be discussed later (Chapters Ten and Eleven) it is necessary to balance the economies of each of the factors of production. It might, for example, be more economic to use a smaller and less efficient machine for the operation, pushing up prime costs, in order to have a machine which could be more intensively used on the site and in the firm, reducing plant overheads by more than prime costs were increased.

Overheads generally raise as the scale of the operation and firm rises. The economies of the scale of management are generally much less than of plant. As the scale of operations is increased more managerial and supervisory staff are generally required. These additional costs may be worth incurring if as a result prime costs can be reduced, for example, through better operational management, lower prices for materials or cheaper credit (Chapters Eleven, Thirteen and Fourteen).

Land development provides an interesting example of the operation of increasing and decreasing returns. The total costs of providing a dwelling declines as the number on a given site is increased as long as no change is made in the form in which the dwellings are provided. If the limit in the number of houses which can be accommodated on the site is exceeded and high flatted blocks are used to achieve a higher density, the construction costs of a dwelling rise faster than the decline in land costs per unit, so that total costs per dwelling rise.

TYPES OF COSTS

It is necessary to distinguish between total, average and marginal costs. Total costs will generally increase with the scale of output. If returns are increasing total costs will be increasing less than in proportion to increases in output and hence average costs will be falling. Marginal costs, the cost of the final unit produced will start to rise when the economies of scale have been exhausted, although for a time average costs may not rise.

DEMAND AND SUPPLY CURVES

Generally the price customers are prepared to pay, the demand price, tends to fall with the volume available. This is because marginal satisfactions per unit acquired tend to fall. For example, a single hammer is necessary to a carpenter, two hammers might be a convenience, each further hammer would tend to be less useful. If the price was thought to be high he would equip himself with only one hammer, at a lower price he might purchase two; the lower the price the greater the purchase would tend to be. The demand price for all the units purchased at the same time is the price of the commodity only just worth purchasing. The total quantity sold depends on the price asked. Normally this would not be less than the marginal cost of production. The best returns would be made when the marginal cost of production were at a minimum and also equaled the demand price at that level of output (Fig. 2.1) If a higher price were demanded less would be sold and marginal costs would be higher. If more were put on the market the offer price would tend to be less and marginal costs higher. Except where the seller has a monopoly he cannot limit supply and must take the demand price as fixed and will hence try to produce and sell the quantity for which his marginal costs are at a minimum.

In the above discussion long time costs and prices have been discussed.

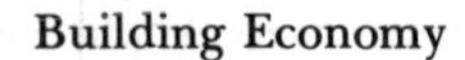

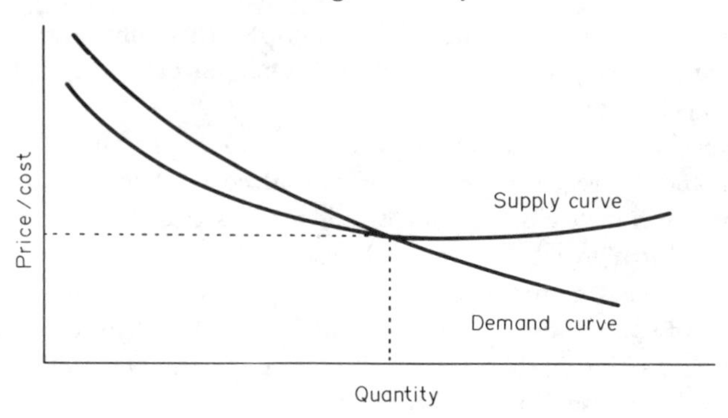

FIG. 2.1 Demand and supply curves

Costs have been taken to include normal profit, one that makes it just worthwhile to incur the risks of production. Prices will tend to be greater or less when demand and supply are out of balance. If demand increases purchasers will be prepared to pay a higher price and prices will be higher at each level of demand. If supply costs have not increased higher profits will be obtained; this will stimulate additional production until marginal costs and prices balance at a new level of supply. Similarly if demand decreases prices at each level of output will fall, profits will be reduced; eventually supply will fall until a new equilibrium is found.

PRIME COSTS AND OVERHEADS

As indicated earlier, costs can be divided into prime costs and overheads. Prime costs are broadly the costs of labour and materials, and are incurred only when a commodity is produced. Overheads, at least in the short run are fixed; they include management, plant and credit. Since in the short run overheads cannot be avoided, it pays to produce at a price which covers prime costs and makes some contribution to overheads. Hence when demand for building falls contractors tender at prices below total costs. In the long run building capacity is reduced, so that when demand increases it tends to exceed the capacity of the industry; contractors' order books fill-up, they become less anxious to secure particular contracts and quote prices which allow above normal profits.

ADJUSTING CAPACITY

The speed with which demand and supply adjust to each other depends on the nature of the industry. A market selling perishable goods has a supply fixed for that day and tends to adjust prices to clear the supply. From one day to another it can often change the level of supply in relation to the previous days demand but it may take a year or more to effect any large-scale change in supply. At one extreme supply may be more or less fixed, for example, old masters, while at the other supply may be flexible, for example, unskilled labour which can find employment in many different fields. Skilled and professional labour in building tends to take a long time to train, most plants for winning and producing building materials take several years to develop, so that while small building firms can be formed quite rapidly, it tends to take a long time to create new capacity once the industry has been run-down. Similarly when demand falls there is generally surplus capacity for some time until the effects of labour drifting from the industry, the non-replacement of plant, less training and the closing of marginal firms and production units work through the industry (Chapters Five and Twenty)

Of course, the construction industry does not produce a uniform product. Resources can be switched within limits from one type of output to another, from building to civil engineering and from new to maintenance work. Often, however, demand changes in a similar direction for all the products of the industry and changes in demand cannot be cushioned by changing the mix of services provided. Producing for stock has limited relevance in this industry partly because much of the work arises on the customer's site and to his instructions and partly because the cost of carrying stock, speculative building, is very expensive except when rates of interest are very low.

JOINT PRODUCTION

Sometimes products are joint; one product cannot be produced without others. Examples would be the winning of sand and gravel and the production of waste from burning coal. Clearly where joint products are produced it will be the ones in highest demand which will contribute most to costs. Prices for commodities with little demand in relation to supply such as fly ash and clinker will be low.

EXTERNAL COSTS

It is necessary to distinguish between costs to persons, firms and public bodies and costs to the community as a whole. In reaching a decision the developer or user generally takes into account only those costs which he has to meet and ignores other costs, external costs which are borne by other groups of people. For example, a manufacturer of cement would generally locate his plant at a location at which costs would be lowest to him and would not take into account costs to others arising from noise, dust and vehicle traffic. Road contractors can often reduce costs to them by closing lanes and do not take into account the congestion costs which this creates for road users. Again the most convenient siting for a building may take light and create noise, dust and other nuisances for the users of adjacent buildings. It is the difficulty of transferring such external costs to developers and users of buildings that justifies the imposition of controls to protect other memebers of the community (Chapter 21)

PRICE AND INCOME ELASTICITY

It will be appreciated that in this chapter it is only possible to mention a few of the simpler economic concepts and only in the most general terms. For example, mention has been made of the way demand price tends to fall as the quantity available increases. This can be expressed more precisely through the concept of price elasticity; this expresses the percentage change in the quantity demand in relation to the percentage change in price. Demand is said to be elastic if the quantity demand increases more than in proportion to a fall in price and inelastic if demand increases less that in proportion to a fall in price. A measure of the elasticity for a commodity can be estimated from statistics of past levels of price and demand. Demand is generally elastic if a commodity has close substitutes, for example, a flat is a close substitute for a house, and inelastic if there are no close substitutes, as there are not for dwellings.

In a similar way the concept of income elasticity expresses changes in demand relative to changes in income. Again measures of elasticity can be estimated and used to guide market decisions. Generally in the case of necessities demand increases less than in proportion to changes in income and more than in proportion in the case of luxuries. Income elasticity of demand would be expected to be higher for housing improvements than for basic housing.

Total demand for a commodity depends not only on individual incomes but on the total number of incomes. Similarly total national demand depends on total national income. Additional economic activity generates a far greater increase in national economic activity when its effects have worked through the economic system. If additional economic projects are started an additional demand is created for materials, plant and labour, additional employment is created and additional incomes. These create yet more demand, employment and incomes and hence yet further demand and so on. This is known as the multiplier effect. Governments often try to stimulate additional employment during trade recessions by ordering new public works such as roads, railways and other constructions. It will be appreciated from earlier discussion that if orders are created beyond the current capacity of industry or faster than additional capacity can be created, prices will tend to rise, imports increase and other orders choked off (Chapter Five).

These illustrations of economic concepts and principles will have indicated that they are basicly commonsense and are intuitively applied in reaching industrial, design and business decisions. The value of a formal approach lies in the deeper understanding which formal analysis provides and in the analytical tools which can be created to guide decision making. Some of these will be illustrated in later chapters.

CHAPTER THREE

BUILDING PRICES AND NATIONAL EXPENDITURE

TRENDS IN BUILDING PRICES

Historical evidence suggests that the costs of buildings will tend to rise relatively faster than costs in general. This means that the real costs of buildings, that is the costs after discounting the effect of changes in the value of money, will tend to rise relatively to the real costs of other things.

There is little doubt that, over the last few decades at least, buildings have been rising in quality. Generally, at least in the Western countries, there has been a rise in space standards, in the quality of components and in the standard and provision of fittings, particularly for engineering, plumbing and electrical fittings. As a result a better product has been provided but generally this has cost more. This increase in standards is sometimes forgotton when comments are made about the increase in the cost of buildings.

There are some grounds for expecting that, in the long run, quality apart, the prices of buildings, unit for unit, will tend to increase in relation to other things. As pointed out earlier, building is a very ancient industry and in many ways, at least in Western countries, neither the product, nor its production has changed very much over a long period, although new forms have been developed and there have been changes and extensions of fixtures and fittings, especially for engineering services. Unit costs in real terms also tend to rise with labour unit costs, that is in marked contrast to conditions in new industries in which prices tend to fall rapidly as better methods of production and organisation are found and advantages are obtained from large-scale production. Compare, for example, the conditions of the motor industry in Western countries in the early 1900's, or the radio industry in the 1920's with the house-building industry of those periods. Since those times the car and radio industries have changed from the production of hand-made products on a small scale to mass production with a high degree of division and specialisation of labour and very large markets. In contrast, the scale of house-building has hardly changed; houses are still built largely by hand and with a largely unchanged form of organisation. Whether this situation is sensible will be a matter for later chapters. In the circumstances it is not suprising to find that the real prices of cars and radios have fallen dramatically in relation to the real prices of dwellings. In fact, there are many new industries which have been developed over a short period from small-scale, hand-made production to highly mechanised mass production. It is not, therefore, suprising to find that building prices have risen disproportionately to the prices of other goods and services over the last few decades.

The measurement of price movements of different types of goods has become only generally possible in recent times and it is difficult to find statistical evidence for more than a few countries from which to measure historical price changes. Price indexes indicate that in America, for example, building prices increased over twice as much as the prices of consumer goods during the period 1913 to 1960.(4) Over the same period, building prices appear to have increased half as much again as consumer prices in Sweden(5) and three-quarters as much again in Great Britain.(6) Over broadly the same period house-building prices have increased three-quarters as much more than wholesale prices in France and have more than doubled in Holland over the half century.(7) Over the thirty years to 1975, house-building costs rose seven and half times, house prices rose seven and a half to eight and a half times and general prices five times.(8) The difference in the ratios seems to reflect the greater increase in productivity in general industry as compared with the building industry. It is, however, notable that the greatest rise in the cost of building relative to the prices of consumer goods appears to be in America, which is generally acknowledged to have made considerable advances in industrial productivity.

During short periods, of course, the movement in building and other prices may be the reverse of the long-term trend. This might happen, for example, during a period at the start of which productivity was especially low, for example, in Britain in the nineteen fifties over the period 1958 to 1968 construction prices rose only 93 percent as much as retail prices. This situation was reversed over the period 1969-80, when construction prices rose a third more than retail prices (Fig. 3.1). During the sixties and early seventies construction prices rose faster than consumer prices in most European countries.

TABLE 3.1

MOVEMENTS IN CONSTRUCTION PRICES RELATIVE TO THE PRICES OF CONSUMER GOODS

	Price Movement, 1960-74
Belgium	131
France	108
Ireland	127
Italy	191
Holland	138
United kingdom	121
Western Germany	136

SOURCES: EEC Publications

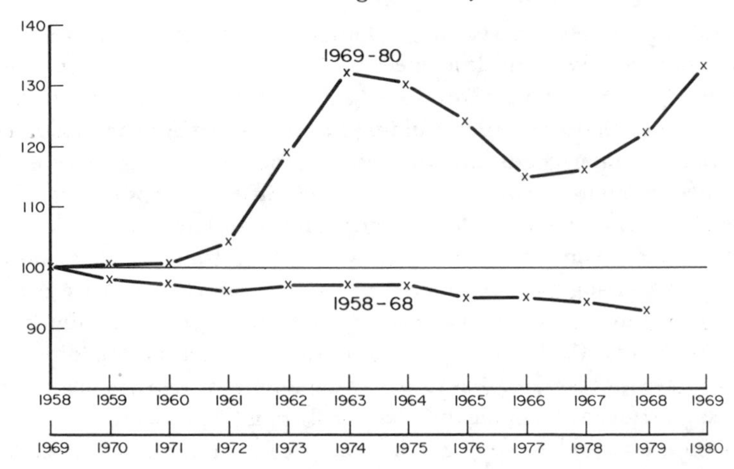

FIG. 3.1 Movements in prices for construction as compared with retail prices

If building prices rise disproportionately to other prices the community will need to spend a larger proportion of its resources on building and this will be reinforced if the buildings demanded are built to a higher standard with more fittings and floor space.

CONSTRUCTION AND NATIONAL RESOURCES

There is nothing sacrosanct about the proportions of national or personal income to be spent on buildings providing that sufficient remains for the other essentials of living. If productivity rises in other industries a smaller proportion of resources will be needed to maintain the same output of general goods and a larger share of the resources will be available for building.

Nevertheless, it is likely in most circumstances that the greater the relative prices for building services, the lower the effective demand for them, since the amount of satisfaction relative to the price paid will fall relative to that provided by other goods and services. It is, therefore, in the interests of the building industry, as well as the community, for the building industry to be as efficient as possible. The stock of buildings is large compared with the annual production and so sudden changes in the price of buildings could result in large flunctuations in demand.

PRICES OF LABOUR AND MATERIALS

Generally, the long term movement has been for the prices of building materials to move up less rapidly than labour rates, or even to fall while labour rates rise. In America, over the period 1913 to the early sixties, labour rates increased ten-fold while material prices rose only four- to five-fold.(8) In Sweden over the same period there was an increase of about six-fold in labour rates and about five-fold in materials.(5) The corresponding figures for Great Britain were perhaps something over a six-fold increase for labour rates and four- to five-fold for materials.(6.9) It is interesting to note that in Great Britain in the 60 odd years prior to 1913 wage rates doubled and material prices fell about a sixth. Again, in the last decade labour rates have usually increased more than the prices of materials. Since in most cases it has proved easier to mechanise material production than site production, it is likely that this trend will continue. If the prices of buildings are not to rise, the increases in labour rates must be offset by increases in labour productivity. It will, however, be appreciated that raising labour productivity has itself a cost in terms of machinery, supervision, organisation and incentives; at times the cost of these has largely offset savings resulting from increased productivity. Savings resulting from more efficient design are not lost in this way.

CONSTRUCTION ACTIVITY

The real output of the construction industry in Great Britain increased in the nineteen sixties and early 'seventies at a substantial rate but declined equally rapidly in the late 'seventies (Fig. 3.2). Generally new work is about two-thirds of total work but was a higher proportion during the late 'sixties and early 'seventies. Much maintenance work cannot be delayed however difficult the financial circumstances and hence maintenance work tends to be less volatile than new work. New housing work for private purchase depends on the availability of mortgage finance and the relation between incomes and the annual costs of housing, including interest. Other private building depends on the buoyancy of the economy and business confidence.

FIG. 3.2 Changes in level of construction output. Great Britain

Public building depends to a large extent on government economic policy.

Construction activity tends to be volitile in most countries. Even where the trend is rising, the rise is usually uneven (Fig. 3.3).

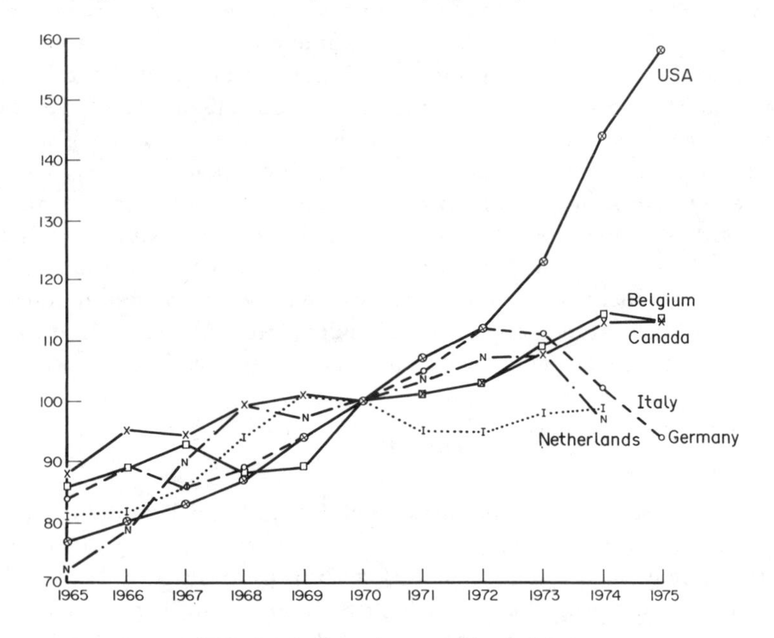

FIG. 3.3 Indices of construction activity

To a large extent construction varies with movements in the economy so that the proportion of national output accounted for by construction activity is far less variable than construction activity itself. However, the importance of construction activity relative to other activities can itself vary. In the fifties and 'sixties the relative importance of construction rose following the war. In Great Britain it rose from under 9 per cent of national product to a peak of 17 per cent and then leveled off at about 12 to 13 per cent, about an eighth. Thus the construction industry as interpretated in this book, that is including materials production as well as contracting and professional work is a major element in the economy and makes a large contribution to national capital formation, that is to the national stock of capital.

However, in many countries the demand for construction work is considerably modified by government action. In the commercial and industrial sectors government economic, financial and fiscal policy often affect both the need for construction and the ability of firms to finance it. Government may affect the demand for private housing both through such policies and through social policies, such as rent control and housing subsidies. The public sector is generally directly responsible for a large proportion of orders for construction work such as public housing, public buildings and infrastructure. Together these often amount to a half of the orders for construction work. Public sector demand is less influenced by prices than private sector demand. While their demand is related to conceived public need, in the shorter run it is generally more affected by economic and financial policy. Capital Expenditure, and particularly expenditure on construction, tends to be cut back when economic activity declines, cutting construction demand offering an easy way of cutting public expenditure. This tends to lead to backlogs of essential work. Governments tend to increase the demand for construction work when they wish to stimulate the economy.

CHAPTER FOUR

BUILDING PRICES AND THE CLIENT

THE MARKET FOR BUILDINGS

While there are a few luxury goods whose high prices create their value, buildings do not fall, by and large, into this class. Generally the purchasers of buildings require the greatest utility at the lowest price—the best value for money.

It is convenient to divide clients into two main classes: clients for whom buildings are a means of production or an investment, and clients for whom buildings are an end in themselves. The first type of client requires buildings as a part of the process of some market activity, the building is a factor of production and has a calculable demand price. The second type of client requires the building for consumption in the form of the amenity which it offers over its lifetime. People both individually and collectively belong sometimes to one class and sometimes to the other.

Generally, commercial and industrial building users, building financiers and investors belong to the first class. Public authorities are also in this class in so far as they build for letting to others, or for their own market services. The situation is not changed when the buildings are subsidised, since the aim is still to balance costs and revenue at some level.

Householders and public and private bodies who provide services without charge fall into the second group. Such building users require buildings for their utility and not as a means of production. Their demand prices are related to what they can or are prepared to afford out of income or capital rather than the price worth paying for a factor of production.

There can, of course, often be no clear-cut division between those who require buildings as factors of production and those who want them for their utility. Business organisations often give some thought to the building's appearance over and above its value for business purposes, and individuals often give some thought to the value the building will have on re-sale.

Up to a point, as the price of a commodity falls, demand tends to increase. Since buildings, even of the same type and serving the same function, vary considerably in quality and size, a fall in price tends to be reflected in a demand for higher quality and larger buildings as well as for more units of building. Most buildings have a long life and hence the stock of buildings tends to be large compared with the annual production. Building users who can only afford to pay low prices either purchase small, low-quality, new buildings, or purchase or rent old buildings. As buildings become older they tend to be less well fitted to meet current needs and hence command lower prices. There is thus a gradation of buildings of different ages and offering different utilities, and, of course, commanding different prices. People with more limited resources purchase these lower-price buildings. Producers, unable to afford suitable modern buildings, make do with old buildings adapted to their purpose. In this way the economic lives of buildings are extended and the needs of those unable to afford the price of first-class buildings are met.

Because, in a competitive market, the prices of new and old buildings are related, an increase in the price of building affects the prices of all the buildings, at least all those for which contracts are about to be made or renewed. Thus if the price of building rises relatively to other things some people who would have required new buildings will accept existing buildings. While, collectively, more will be spent on accommodation there will be a tendency for building users to accept less utility from buildings; some buildings will not be built, and some which would have been demolished will remain in use. Since the stock of buildings is large compared with the annual production an increase in price will tend to result in a disproportionate fall in demand for new buildings. This is, of course, one of the reasons why building construction sometimes fluctuates violently from year to year.

DURABILITY AND RATES OF INTEREST

Because buildings have a long life and are expensive both in relation to householders' income and to the annual turnover of most businesses, it is normal to purchase buildings with borrowed money. The renting of a building is really only borrowing the capital in perpetuity. In fact, the difference in cost between a loan in perpetuity and a long-term loan is marginal. Thus the effective price for a building is not generally its capital cost but the cost of servicing the loan for its purchase.

As will be explained later (Part 4) the price of the structure itself is only a part of its total cost. To it must be added the equivalent costs of servicing the structure, that is, maintenance, heating, ventilating, cleaning and other costs of providing the required internal and external environment. The situation is not substantially changed when the building owner uses his own capital, since by doing so he loses the income he would otherwise enjoy. Hence the effective price of buildings is affected by the rate of interest prevailing as well as by the costs of construction. The annual repayments are more or less doubled over the range of interest rates from 3 per cent to 7 per cent and doubled again over the range from 7 to 15 per cent. Generally, the range for any one place and over a medium period of time is about 2 per cent, but this can make a difference of 40 per cent in the annual repayment (Table 4.1). It will be noticed, for example, that an increase in the rate of interest from 4 to 5 per cent has

TABLE 4.1

CAPITAL AND RATES OF INTEREST
ANNUAL REPAYMENT OVER 60 YEARS

Units of Money

Capital Cost	Rate of interest									
	3%	4%	5%	6%	7%	8%	9%	10%	12%	15%
800	29	35	42	50	57	65	72	80	96	120
900	33	40	48	56	64	73	81	90	108	135
1000	36	44	53	62	71	81	91	100	120	150
1100	40	49	58	68	78	89	100	110	132	165
1200	43	53	64	74	85	97	109	120	144	180

about the same effect on the annual repayment as an increase in price of 20 per cent. As will be seen from Table 3.1 these results are based on a loan period of 60 years. A corresponding increase in the rate of interest for a loan period of 20 years increases the annual repayment by less than 10 per cent (Table 4.2). Clearly, the shorter the term the less the influence of the rate of interest. At the same time the relative increase in annual payments falls as the rate of interest increases.

But in practice the effect of the rate of interest on the cost of buildings is modified in two ways: by taxation, and by inflation.

TABLE 4.2

CAPITAL AND RATES OF INTEREST
ANNUAL REPAYMENT OVER 20 YEARS

Units of Money

Capital cost	Rate of interest									
	3%	4%	5%	6%	7%	8%	9%	10%	12%	15%
800	54	59	64	70	76	82	88	94	107	128
900	60	66	72	78	85	92	99	106	121	144
1000	67	74	80	87	94	102	110	117	134	160
1100	74	81	88	96	104	112	120	129	147	176
1200	81	88	96	105	113	122	131	141	161	192

THE EFFECT OF TAXATION

Normally, at least in Western societies, private purchasers and business firms are tax payers. While the tax regulations differ from one country to another their effect is often to reduce the effective cost of a building to the building user. In the United Kingdom the effect of taxation is virtually to halve the cost of servicing the capital, whether the firms or individuals use their own or borrow. This situation arises because interest is treated as revenue or income for tax purposes. Hence taxable income is reduced if interest is foregone or paid. The effects of taxation are often much less for house purchasers. The rent of a business property counts as a business expense and so, again, reduces revenue and the effective cost of the building is again about halved. There is, however, often no corresponding relief for householders paying rents, although public authority housing is often subsidised. For example, in Britain the amount of the subsidy is generally about equivalent for comparable dwellings to the tax rebate which a private householder who was a net borrower would receive on the interest payable on a mortgage. Thus business building users, and up to a point owner-occupiers who are net borrowers receive a substantial tax benefit which may reduce the cost of a building by as much as one half. The investor in buildings pays tax on the income realised as does any other investor. The govenment obtains a tax revenue from buildings rented for private occupation that they do not obtain from other types of occupation. The incidence of taxation varies from one country to another according to its tax laws.

THE EFFECT OF INFLATION

In the long term, the value of money tends to fall, the rate at which this happens varying from one period to another and from one country to another. For the decades since 1945 inflation has been general in most countries and the value of money has fallen considerably. For example, in the United Kingdom the general price level rose on average about 3 per cent per year for the first two decades. Since rates of inflation have increased phenomenally, prices have trebled since 1970. Money borrowed to purchase buildings is reckoned in the form of money debts; during a period of inflation its real value falls year by year. If the rate of inflation is 5 per cent a year a £100 loan at the beginning of the year is worth only about £95 at the end of the year—that is, it will purchase only about 95 per cent of the real goods at the end of the year that it would have purchased at the beginning. This, of course, means that the borrower pays 5 per cent less in real terms than he contracted to pay. Thus with inflation at 5 per cent the real cost of borrowing is reduced by this amount and a 8 per cent rate of interest is effectively only 3 per cent. At high rates of inflation the effective real rate of interest may be negative, as it has been at times in several countries in the past few years.

THE DEVELOPER

The price a developer can afford to pay for a building is often closely circumscribed. On the one hand, the rent or the sale value of the finished building is determined by the market in relation to the size and quality of the building and its position. On the other hand, the price offered for a site is determined by the market in relation to its position and current demand. The developer must meet the costs of construction, fees, administration and finance, and secure his profit from the difference between these two figures. Since the stock of buildings is large compared with the annual amount of new building, the market price for new buildings is determined more by the demand than by the costs of construction. The price of a site must be sufficient to make it worthwhile for its owner to move and to carry on his activities elsewhere.

There is, of course, a relationship between the size and quality of a building and the price it will fetch. Hence, up to a point, it is worth spending more on a building if, as a result, it will fetch a higher price. The expected price of a building therefore implies a given size and standard; the development will be worthwhile only if it can be built for the price allowed by the investment. For example, suppose the current price in a particular district for office accommodation is £150 per square meter of rentable floor space, inclusive of maintenance, heating, cleaning and other services. If the building and planning regulations allowed a building which would provide 200 square meters of lettable floor space, the revenue would be £300,000 a year. Suppose the annual costs of the services and management, including an allowance for profit, were £140,000, this would leave £160,000 a year to be capitalised to meet the costs of the building project (Table 4.3). If money could be borrowed at about 7.5 per cent net the capitalised value over a 40 year period would be about £2,000,000. If the site cost £550,000, and fees, expenses, developer's profits and contingencies about £650,000, this would leave £800,000 to meet the cost of constructing the building, that is about £400 per square meter.

The period taken for capitalisation was 40 years, it being assumed that the building would continue to provide the assumed revenue over this period. Suppose, however, that the site could only be obtained for a short period, or that the building was only expected to be needed for a short period, for example, 20 years. Suppose also that the net rate for borrowing was about 8 per cent net. In this situation the capitalised value would only be about £1,600,000, clearly insufficient to meet the costs of development. If the costs of the development were as given (Table 4.3) the rent would need to be about £170 per square metre to make the development worthwhile.

Clearly, short-life buildings inevitably have higher rents than those with long lives. A change to shorter life spans for buildings would lead to higher rents and a fall in demand for building space unless the costs of construction

TABLE 4.3

THE FINANCE BALANCE SHEET FOR A BUILDING DEVELOPMENT

Revenue 2000 square meters at £150 per square meter		£300,000
Equivalent annual expenses for maintaining, cleaning, heating and generally servicing the building	£100,000	
Management and profits	40,000	
Balance to be capitalised as cost of development	160,000	
	300,000	
Capitalised value of $160,000 at net mortgage rates of 7.5%		£2,000,000
Cost of site	£550,000	
Professional fees and expenses	150,000	
Developer's profit and contingencies	500,000	
Amount available for construction site clearance, etc.	800,000	
	2,000,000	

could be reduced. Similarly, the owner-occupier would not be prepared to pay as much to purchase a building if it had only a short life since the purchase is really only the capitalisation of the annual rent.

SHORT-LIFE BUILDINGS

The effect the length of life has on the capitalised value of the rent depends on the rate of interest. The capitalised value increases with the length of life but at a declining rate of increase until a point is reached at which the capitalised value is relatively unaffected by further increases in the life. At a rate of 3 per cent interest the capitalised value reaches relative stability with a life of about 70 years, the corresponding life at a rate of interest of 5 per cent being about 60 years, and at a rate of interest of 15 per cent about 40 years. Generally, there is little saving in the costs of construction if buildings are constructed to last shorter periods than 60-70 years, so that shortening the lives of buildings reduces their value without reducing their costs. In fact, if a building is to last more than a few months a form of construction will need to be used which will generally last 60-100 years. There are, of course, some forms of construction which, if not properly maintained, will deteriorate in a few years to a point where they cannot be used; the rate of deterioration will depend on the climatic conditions. Maintenance is generally necessary to ensure that the fittings function properly, that the building remains weather-tight and of acceptable appearance. Generally, the neglect of maintenance results in a loss of value greater than the saving in costs.

There are many situations for which short-life buildings would be a convenience if they could be produced cheaply enough—for example, for purposes for which changes in technique necessitate frequent changes in building layout and design, and for countries in which standards and needs are changing rapidly. However, it will be seen (Table 4.1 and 4.2) that if, for example, buildings were to have a life limited to 20 years, with rents no greater than for long-life buildings, with moderate rates of interest it would be necessary to be able to construct such buildings for a half- to three-quarters of the cost of long-life buildings, the ratio depending on the rate of interest. The higher the rate of interest the less drastic need be the reduction in the construction costs. With interest at 5 per cent the ratio should not exceed about two-thirds, at 3 per cent the ratio should not exceed about a half, while at 7 per cent a ratio of three-quarters would be acceptable.

In the absence of short-life constructions available at equivalent annual costs to those for long-life constructions, it is more economic to accept the need for adaptation of long-life buildings as the need arises. When adaptation is not economic either the building is demolished, or it is used for a less exacting purpose.

CHAPTER FIVE

THE ECONOMY, BUILDING DEMAND AND THE CONSTRUCTION INDUSTRY

FINAL BUILDING USERS

The demand for buildings arises in the final analysis from those who wish to occupy them. Households may purchase dwellings for their use or rent them from private owners, housing associations and increasingly—in Great Britain—from local authorities. Industrial and commercial firms may purchase their own premises, but increasingly they rent from building investors so as to conserve their capital for use in the businesses and because generally this is cheaper when allowance is made for the effect of taxation and the relative costs of raising capital. Public authorities, especially those concerned with production or trading, often rent their buildings. Buildings required for providing a community service are generally owned by the public authority.

PRIVATE HOUSING

Generally with the growth of rent controls and the provision of housing for rent by public authorities, only luxury dwellings are built for rent by private investors. Most private housing is owner-occupied. However, few owners have their dwellings purpose built, most purchase new dwellings from housing developers and existing housing from their current owners.

Generally the finance for the purchase of housing is borrowed; in Great Britain, until recently about five-sixths of it was borrowed from building societies, the balance from insurance offices, banks and local government. Recently banks have extended their lending for house purchase. The amount of finance available for lending for house purchase, still depends to a large extent on the savings invested in building societies. This depends on the cost of living in relation to income and the amount available for saving, on the relative rates of interest paid by building societies to lenders in relation to other outlets for small savings, on taxation and tax expectation and on the rate of inflation and expectations about the rate of inflation. If for any of these reasons the amount invested fails to increase at an adequate rate or declines, building societies have less to lend and can grant mortages to fewer house purchasers and lend less to each. Thus the rate of investment affects both the number of house purchasers and the prices they can afford.

The stock of existing dwellings is large (forty to fifty times) compared with the normal rate of additions to the stock. Even though only a proportion of existing dwellings are on the market at any one time, the number is usually two or three times as great as the number of new dwellings. As a result new construction cannot greatly affect supply in the short to medium run. Hence the prices of dwellings in the short to medium run depend more on the availability of mortgage funds than on building costs, and prices can rise and fall substantially over short periods. Since the amount of the mortgage is limited in relation to the borrower's income, the number of new borrowers declines with an increase in dwelling prices unless incomes rise in step.

Occupiers purchase dwellings not only to consume the amenities they provide but also because dwellings provide a store of value, that is their price tends to increase in step with inflation and thus retains its real value in contrast to savings held in building societies and other savings held in money terms. Thus when the value of money is declining, or is expected to decline, and especially when there is little confidence in equities, people put their savings into property, in the case of housing by purchasing a house, or a larger one, extending an existing one or purchasing a second one. The extent to which they do this depends on expectations about the rate of inflation, the rate of interest and tax regulations, the availability of mortgage funds and incomes in relation to house prices. In Great Britain the ratio between house prices and earnings usually lies between 3 and 4. Such transactions add further to the rate at which the demand for dwellings can change.

Developers cannot respond rapidly to changes in the demand for dwellings for purchase, partly because stock once built remains, partly because the acquisition of sites, the obtaining of planning permissions, the planning, design and organisation of development takes a long time and partly because the banks and other finance houses may not have the money to lend to finance land acquisition and development or not on acceptable terms. When demand for dwellings is buoyant, prices rise and may stimulate further price rises and developers are prepared to pay more for building sites in order to build more dwellings. Such buoyant demand tends to be eventually checked by the rising prices and a short fall in building society and other finance to sustain demand at such prices. Because of the length of the time lags dwellings tend to continue in production at a high rate long after the demand has lost its buoyancy and stocks of unsold dwellings accumulate. Financing these stocks and stocks of sites acquired for further development is expensive, builders become anxious to sell, the price of dwellings and sites fall and the contracting industry cuts back its housing development activities causing unemployment and bankruptcies.

The fluctuations in the rate of private house building can be considerable. In Great Britain the rate dropped by three-fifths over the last decade. The changing market conditions have tended to eliminate many small builders with most private house building concentrated in the hands of a few large-scale developers.

Housing standards tend to rise over time; one aspect of which is the rise in space standards. More important, however, is the rise in demand for more and better engineering services such as heating, hot water, lighting and power points. There also tends to be a demand for more built-in cupboards of a high standard, particularly in kitchens and bedrooms. Change of fashion effects demands for existing dwellings perhaps as much as functional needs. There is also a demand for higher standards of thermal insulation and insulation against external noise. Such demand gives rise to a considerable volume of rehabilitation work ranging from replacing single fittings and services to comprehensive remodelling of dwellings.

Such development proceeds in both the public and the private housing sectors. In the public sector the work is generally carried out by contractors or direct labour, while in the private sector a substantial proportion of the work is carried out by the householders themselves. At the same time a substantial number of specialist contractors have been set-up to meet this demand. Some manufacture the equipment they fit. Unlike most other contractors many of these specialist contractors follow aggressive marketing policies; some have a particularly poor reputation for quality and business practises. Rehabilitation work naturally gives rise to an additional demand for building materials and components. Specialist retailers have tended to set up to supply householders with materials and components in suitable packs for their scale of operations. A market has also developed for plant and tools suitable for the home handyman.

Rehabilitation needs to be distinguished from maintenance, which in part it replaces. In the housing field the occupier increasingly carries out maintenance work. The need for maintenance tends to arise, for all types of buildings, as standards rise and particularly as more and more sophisticated equipment is installed. The distribution of maintenance and rehabilitation work tends to be very uneven. Many buildings, particualrly dwellings, receive inadequate attention; eventually they need either extensive rehabilitation or complete rebuilding.

While the relation between new construction, and maintenance and rehabilitation work varies, usually the latter two constitutes a quarter to a third of the work in the housing field.

COMMERCIAL AND INDUSTRIAL DEVELOPMENT

The demand for commercial and industrial buildings depends on business confidence and on the rate of innovation and change in business activities and location. Firms generally only require new and additional buildings when trade is buoyant and they feel confident enough in the future to re-equip or expand their business. Innovations and changes in business activities tend to result in both a demand for new buildings to meet changing technological requirements, particularly for industry and distribution, and for additional buildings, for example as the demand for office services increase. Relocation from old to new towns and from one region to another also adds to the demand for new commercial and industrial buildings.

As explained earlier, while some firms own the buildings they occupy, a large proportion rent from property companies and other investors in property. Property is developed for rent when developers are confident of the adequacy of future returns after tax from such investment. The higher the expected future level of rents, and this expectation rises with business confidence and anticipated inflation, the higher the level of costs developers can contemplate meeting and the more they are prepared to pay for suitable sites (Chapter 4). The amount of finance developers can raise either directly in the market or indirectly by selling previous developments also depends on anticipated future returns. Again a lack of confidence in the equity market and low returns tends to result in more finance being available for commercial and industrial development, while the promotion of tax and other legislation against property development reduces it.

Whereas building contractors are generally themselves the developers of private housing for sale, fewer contractors engage directly in the development of commercial and industrial building but rather are employed by the developers. The developers depend for finance partly on share capital, partly on medium-term fixed interest, finance from investment houses such as insurance offices, pension funds, financial trusts and merchant banks, and partly on short-term finance from other banks. The availability of medium-term fixed interest finance has tended to decline when rent expectation and hence profits have risen and its price has tended to rise. In these circumstances insurance offices, pension funds and other lenders have insisted on an equity interest and some have become developers in their own right. Increasing activity in this field results in increasing demands on the building contracting industry. The buoyancy of such demands depends, however, on a sustained demand for new commercial and industrial buildings, on rising rent expectations, adequate and cheap credit in relation to rates of inflation and on favourable taxation. If these expectations are not fulfilled, development activity is likely to be reduced and with it demand on the construction industry.

Rent expectations decline not only when a fall in economic buoyancy reduces relative or absolute growth in demand for commercial and industrial buildings but also when government restricts rent increases. On the contrary, controls on development usually result in rises in rent levels. The price of credit for industrial and commercial development rises and its supply falls if credit becomes generally restricted as a result of government policy, if confidence is reduced in rising rent expectations, or if relative confidence in other forms of investment increases. Changes in taxation may reduce the net returns from property investment in relation to other investment or in relation to the rate of inflation. In the long run construction will decline as a reuslt of the operation of such forces, until relative shortages of buildings results in rises in rents and other changes, and development once again becomes profitable.

Commercial and industrial buildings also give rise to rehabilitation work and to maintenance. This represents about a fifth to a quarter of the construction work of this sector.

PUBLIC DEVELOPMENT

Public development is carried out by a number of public agencies. In the United Kingdom the central government directly orders and finances the development of offices, hospitals and other buildings for its own use and such infrastructure as trunk roads, and in various ways controls the expenditure of nationalised industries, public utilities and local authorities. Each of these types of organisation develop buildings and infrastructure for their own needs, while local authorities also develop housing for letting—in Great Britain they supply about a third of the dwellings. Thus in the United Kingdom about half the work of the construction industry is subject to close government control. A high degree of government control is normal in most European countries, the form of control varying with the way relations are organised with central government.

The government's control over demand in the public sector operates in a number of indirect ways as well as directly though contracting construction work. In the United Kingdom the central government usually budgets for given amounts of capital expenditure for various purposes. It can control the expenditure of the public agencies through limits on the amount they can borrow through government agencies and from the money market, through limits on capital expenditure in different fields; for example, on road development, public utilities, schools and housing, through capital grants for development and through revenue grants which are necessary to agencies, such as local authorities both to finance debt servicing and to operate the facilities created. Governments also operate other forms of control on construction which relate to forms and standards of facilities (Chapter 12).

In addition to the controls operated by central government to influence the level of construction orders placed by public agencies, the agencies' ability to place orders for construction are also affected by other aspects of government policy. For example, government action which limits the availabiltiy of finance in the market and changes its price, or which stimulates or limits the rate of inflation, or controls the prices that can be charged for the goods or services produced by the agencies, or the ability of firms and households to meet prices and local taxation; all affect the amount of development which public agencies can afford to undertake.

Public agencies also carry out extensive rehabilitation work, especially on housing, and maintenance work. Such work accounts for about a third of the construction work for the public sector. Generally it is less easy to postpone maintenance work than new work, except over short periods. In the long run either the backlog of work must be made good, or the buildings or works will need to be replaced.

DEMAND ON THE CONSTRUCTION INDUSTRY

It is clear that indirectly in the case of private agencies and more directly in the case of public agencies, central governments are able to have a considerable influence on the level of demand on the construction industry. Governments influence not just the level of demand in total but also the level of demand for the various types of work the construction industry undertakes.

The construction industry is very heterogeneous. It consists of numerous firms of a great range of sizes, carrying out different mixes of work. Most firms specialise to some extent or at least are geared to carrying out given classes of work. Since each class tends to require a particular mix of operatives with particular skills, appropriate plant and management expertise, construction firms have only a limited ability to switch from one mix of work to another. Moreover, while some firms operate on a national basis, many do not operate outside their region and most small firms operate only within their own locality. Hence the extent to which changes in demand can be spread across the industry are limited. Only in some limited fields such as road construction and other civil engineering is labour mobile. Mobility is expensive since labour working away from home has to be paid accommodation allowances.

Generally as demand for a particular type of work increases, tender prices rise, contracts take longer to carry out and completion dates become further and further ahead. As the available slack and unemployed resources are used up, productivity often falls as less experienced and well-qualified labour and management are used. Since some crafts are common to other industries as well as construction, a sufficient high price will draw some labour from other industries, but without experience of the site conditions it may be less useful than experienced site labour. Rising demand tends to run into material shortages as well as labour, management and equipment shortages. Of course, in the long run the construction industry can be expanded, but the demand pressure may not be sustained for a sufficiently long period to achieve this. While non-craft labour can be trained on the job in a fairly short time, many crafts require a long apprenticeship. Young men will not come forth in sufficient numbers and contractors will not be prepared to operate training schemes if there is no confidence that the level of demand will be maintained over a long period. Similar difficulties will arise in the training of management. After the slack has been taken up, manufacturers of materials and components can only meet further increases in demand by building additional plant. This will take some years and again will not be undertaken without confidence in the level of demand being sustained.

When demand declines, tender prices fall and unemployed resources increase as current orders are completed. Some contractors who have only small resources go bankrupt as the money needed to repay past debts exceeds current payments. Labour is laid off, some workers retire and some are absorbed by other industries, training is cut back and fewer enter the industry. This pattern is repeated in a modified form in the professions. Materials producers increase stocks before cutting production. The less efficient plants are closed and the building of modern plants is delayed. The longer the period of declining and low level demand, the less able the construction and material producing industries are to meet demand when it again increases.

The greater the difference in demand at the top and bottom of the cycle and the more frequent these are, the less well adapted the industries tend to become to meet higher levels of demand and the higher tend to be the tender prices and the slower the rate of completion.

The amplitude of the cycles of construction demand tends to be greater than those for most other industries. This is because new work and even to some extent maintenance work can usually be postponed in times of financial difficulty. Except where population is increasing or migrating very rapidly, or substantial technological or other changes create a demand for new development, postponement of new work only involves delaying the introduction of new standards; except for actual repairs, delays in maintenance again only involves a postponement of a rise in standards. Businesses are reluctant to invest when there is little confidence in the future and hence postpone the development of new buildings and improvements. Private individuals generally have less money and less confidence to spend when the economy has lost its buoyancy and employment is less certain. Thus a declining economy is likely to be acompanied by a fall in private orders to the construction industry, but a fall much larger in amplitude than in the fall in an industry that produces consumption goods and services, since consumption is necessary to sustain living.

Governments can and usually do similarly cut back demand for construction when a declining economy reduces the buoyancy of taxation, or when the economy is considered to be over-buoyant and prices to be moving out of control.

GOVERNMENT POLICY AND CONSTRUCTION DEMAND

It is often advocated that construction should be used by governments to assist in stabilising the economy. On this argument construction demand should be increased when an economy is losing its buoyancy to revive economic activity through the multiplier effect and reduced when additional activity is merely resulting in inflation. Inflation can arise not only because fully employed resources are bid up by competing users but because monopolistic suppliers can withhold goods and services from the market until prices rise even though supply exceeds demand. Prices have been pushed up in this way for oil by OPEC and by trade unions. The power to do this remains as long as the monopolistic agencies are prepared to have some of their potential supply unused. Such conditions may induce governments to attempt to reduce the rate of inflation through financial or fiscal methods; even though this tends to reduce the level of economic activity. One such method is to reduce government expenditure. Generally they have little success in increasing their own productivity and find it easier to reduce their services, particularly through reducing capital expenditure and with it the demand for construction work.

As indicated earlier, the government is in a strong position to effect orders for construction. About half the orders are often within the public sector and depend directly on government departments or while within the control of other public agencies are subject to government control for loan sanction, loan finance or revenue finance to service public debt. Private orders can be controlled directly through physical controls, licences or permits, or indirectly through credit control, market operations which affect the rate of interest or the amount of finance, and taxation. Government control of the private sector is, however, negative; it can effectively reduce orders but can only stimulate and not create orders.

The two major problems which arise in using the construction industry as a means of controlling the economy are first the scale of demand change which should be sought and secondly the timing of changes in demand. As argued earlier, demand can be cut by a much greater percentage than is practical for most other industries and this can result in considerable waste and inefficiency in the industry and in the long run lead towards a decline. The problem of timing is also difficult. It can be several years between the conception and completion of a building project. A site has to be assembled and a building designed, tenders have to be requested and received. This alone can take several years. If the construction work is postponed, much of the planning and design work may need to be revised for later use and land may be lying idle, but the money tied-up in the land and in planning and design work must be serviced. There would be a reluctance to carry work this far if there was doubt about its completion. Construction can also take several years. As a result, even if orders were available, a change in the value of orders would take several years to have their full affect on the level of current work. For example, if orders were halved, a high rate of change, and the average contract period was two years, current work would only be cut at most by a quarter during the next year. There would be a similar change in the third year, even if there were a return to the previous level of orders in the second year. The longer the average period of contract, the smaller the effect of any change. But, in fact, the effect is likely to be even slower than indicated. Contracts cannot usually be suspended immediately. The current stage of the work must be brought to an orderly end or the ultimate costs may be raised considerably. Generally the effect in the private sector will be slower than in the public sector, since it will usually pay the private developer to complete the development so that he can sell it or obtain a revenue from it.

Given the undesirable effects of changes in the level of demand on the construction industry, it may be asked whether despite the convenience to the government of using the industry as a regulator, it is desirable to do so. Stable demand is clearly an advantage to any industry and to its clients. Stability of demand will be matched by an appropriate level of supply with the promotion of efficiency and competitive prices. Ideally the government should, from the point of view of clients, aim to regulate the industry so as to reduce the size and speed of swings in demand, so as to reduce the inefficient use of its resources, the inflation of prices and the unnecessary lengthening of contract periods.

It can also be questioned whether government intervention is successful in its purpose and worth the price the economy has to pay for it. Because of the time scale of the industry's operations, effective intervention needs to be made months and even several years before the correction to the economy is required. This requires that the government can recognise the way economic situations are developing several years ahead. Successive governments have shown themselves very inadequate at forecasting the future of the economy over any length of time. It is an open question whether such economic fine tuning does not cause more difficulties than it corrects.

PART II

The Development of Building Construction

CHAPTER SIX

TRADITIONAL BUILDING CONSTRUCTION TODAY

THE INFLUENCE OF LOCAL MATERIALS AND LABOUR

Traditional building construction is simply the form of building which is normal at some particular time and in some particular place. What is considered traditional varies from one period to another and from one place to another. There is often a different tradition for what has been called monumental from that for domestic building. The wealthy client is not limited to a form of construction based on local labour and materials as is the domestic client. The poorer the local supply of materials and skills the greater the differences there tend to be between the two forms of building. In places where there was a poor supply of durable materials, only the State, the Church and the landowner could afford to import the stone and brick for durable construction, and in such places there are usually few examples of domestic architecture surviving from the past to match the examples of monumental buildings. This is, for example, true in Britain, where domestic buildings do not generally date back more than three or four centuries whereas there are churches and castles which were built eight or nine centuries ago.

Most forms of building have traditionally been based on local materials. This is inevitable when materials are bulky and heavy and there is no cheap form of transport. As building skills improve and the properties of materials are better understood, the materials are fashioned not only to improve the convenience of the buildings but also so as to use the materials themselves more economically, which in itself makes it more feasible to obtain materials from other areas. The use of materials in a more economic fashion is stimulated as the more easily available materials are used up and there is a need to go further for materials and to spend more time in winning them. For example, in stone districts the earlier forms of construction were often based on thick walls built of loosely piled stones which were gathered from the ground. Ways of building walls with less material were developed when it became necessary to dig the stones out of the ground and perhaps to cart them some distance.

Some forms of construction are more labour-consuming than others and some need more skilled labour than others. Self-built construction is usually relatively unskilled and often the labour used for such construction has no economic value, since the construction is carried out at a time when the labour would otherwise be unoccupied. Under such conditions the most economic form of construction is one which is based on materials which can be fashioned without much skill and which are readily to hand. The application of these principles can be observed in the forms of construction once traditional in many parts of Britain. In districts with plenty of suitable timber, houses were timber-framed and often boarded; where timber was not available quite so freely and at times before it was economic to use timber boarding, the space between the frames was filled with a hurdling of sticks and covered with a daub of cheaply available local material, such as clay and lime mixtures. Where more money and hence labour was available the clay was used for making bricks which were themselves used as infilling. In East Anglia, where suitable timber for framing was not everywhere so plentiful, load-bearing walls of unbaked bricks of clay and straw were used. In the West Country, where timber was also short, load-bearing walls were built of cob, a mixture of clay, chalk and straw which was mixed by treading, and built up in layers. In other areas pisé was used, a mixture of dry chalk and marl or clay which is rammed and consolidated between formwork.

Structurally, load-bearing walls of earth are quite feasible for small structures. Providing that the walls are kept dry, they are quite durable, are adequate for load-bearing and provide good insulation. Where such buildings are erected on an adequate plinth and are properly protected with wide eaves and properly maintained, they last as long as most other forms of construction. There are many examples of such buildings which are well over a century old. In some countries, forms of construction not dissimilar to these are used today.

In the north of Britain, stone, formally confined to foundations, came into more general use with the destruction of the forests and a rise in the cost of timber. In the clay areas bricks, at first confined to the construction of chimneys, were later used for infilling and, as timber become more costly, for building load-bearing walls.

As construction became more and more a professional business, as it did as people entered full-time employment instead of part-time husbandry and no longer had the time to build for themselves, economy of labour became important and forms of construction needing a great deal of labour gave place to forms more economic of labour. In Britain the development of coal-mining and the construction of canals and railways made available both coal cheap enough to stimulate an expansion of brickmaking and transport necessary to move bricks cheaply from one part of the country to another. The production of clay tiles similarly increased but these could not compete with the light-weight blue Welsh slate.

As a result of the availability of cheap bricks and slates, and the need to employ paid labour, methods of building based on materials available on the site became uneconomic except in exceptional circumstances. However, even in the 1920's the use of wet or rammed chalk was still not a completely uneconomic form of walling. Where the raw material was available on the spot and transport for bricks was poor the saving in materials offset the extra cost of labour.(10) It will be appreciated that in some economies simple methods based on materials obtainable from the site are quite economic, even although expensive in labour.

INTERNATIONAL DIFFERENCES IN TRADITIONAL CONSTRUCTION

In Britain the brick is particularly cheap and is widely used in all forms of construction. Brick is also widely used in some parts of western Europe, although it is not as economic or as widely used as in Britain. In North America bricks are comparatively expensive, while timber is cheap and is generally used both for farming and for cladding small buildings.(10) Again, timber is cheap and plentiful in Australia and in New Zealand and is widely used for framing and for cladding. South-East Asia is again predominately an area where timber is the typical building material. For example,in Burma and Indonesia most small buildings are of timber and bamboo.(10) This is what would be expected in countries with an abundance of forest products. Timber construction satisfies most of the economic and physical conditions required. It is cheap, easily handled by non-professional constructors, provides a flexible form of construction which will withstand earthquakes without too much damage, and which if necessary can be easily repaired, and is adequate for protection against sun and rain in countries with a hot, humid climate. Light-weight forms of construction such as timber and bamboo are not so suitable in hot, dry climates where the need is not so much for ventilation as for the exclusion of the sun's rays and heated air, and hence where thick walls are an advantage. Moreover, in such climates, for example, India, timber is less freely available and is often wanted as fuel for burning clay for bricks and tiles. Again, in some tropical countries timber cannot be used because of the likelihood of damage by termites.

Often the simple indigenous materials are inflammable and hence unsuitable in large urban areas where buildings are required close together. In urban areas the building users tend to be in full-time employment: not only do they not have time to build their own dwellings but they lack the time to carry out the level of maintenance necessary for durability for many of the forms of construction based on indigenous materials found in Asia and Africa. Hence, as countries become industrialised and urbanised the form of building traditional to the country tends to change to one based on non-inflammable materials; construction becomes largely professionalised and labour economy becomes of importance both in the initial construction and in the maintenance of the buildings. For the smaller type of building masonry is generally used, but for large buildings, and these tend to become of increasing importance, steel and concrete are usually used. Since building materials tend to be expensive to import because of their bulk and weight, and since they are generally comparatively easy to manufacture, they are suitable for home production in devleoping countries.

Where suitable clay combined with adequate cheap fuel is available, some form of brick and other clay products can be made. Some clays contain more combustible materials than others; for some clays it is only necessary to use a tenth of the fuel necessary in other cases. The cheapness of the English fletton brick is to a large extent due to the small amount of fuel needed to fire it. Generally, fuel can be saved by using perforated rather than solid bricks. The cheapest alternative to brick for masonry construction is usually concrete blocks; while the cement for these often has to be imported, some form of aggregate can usually be found locally.

Cement is a fairly simple material to produce given modern plant and "know how". Suitable raw materials are fairly widely distributed, and more and more developing countries are manufacturing cement. For instance, in South-East Asia cement is now produced in Burma, Malaya, Thailand, the Philippines and Taiwan (Formosa), as well as in Japan, and there is some export from some of these countries.(12) Cement is also being produced in various places in Africa and in the West Indies.(13)

The availability of cheap cement provides a possible basis for the production of a range of building materials, quite apart from its use for *in situ* concrete. Concrete blocks can be produced without the aid of much machinery or skilled labour and can be developed as a backyard industry. More sophisticated use can be made of precast concrete to produce posts, columns, beams and other bulding components. Or again, the cement can be used as the basis of an asbestos cement industry for the production of sheets, pipes and other products, although expensive plant and skill must usually be imported for this industry to be developed. The developing countries tend to be moving along these lines, the types of industry depending on the local supplies of raw materials and the skill of the local labour.

There is also a tendency for developing countries to produce simple metal sheets and bars and structural steel and to fabricate window and door frames from steel and aluminium. This generally requires more skilled labour than simple concrete products and, of course, the appropriate supplies of materials.

While developing countries are beginning to reduce their reliance on imported materials, the shortage of trained labour often remains. It is generally accepted that better housing is one of the more important ways of raising the standard of living. It is undertaken to provide a higher level of hygiene, healthy and more comfortable living conditions, and greater security for life and property. For these and other reasons discussed above there is a desire to develop a traditional house of greater permanence than the local tradition often provides. This usually means a building of brick or block, or framed and sheeted, and in either case entails the production of materials formerly largely imported and the use of skilled labour of a new kind.

The speed with which permanent housing suitable for an industrialised urban society can be provided depends on the standards set and on the size and development of the local professional building industry. In many countries the local building industry is not large enough to undertake both the building and civil engineering necessary for an expanding industrialised economy and to provide a great deal of housing. Developing countries are meeting this difficulty partly by bringing in overseas contractors for the larger and more complicated buildings and engineering works, partly by setting their initial standards of housing fairly low, and partly by encouraging self-help building where this is possible. The use of overseas contractors results not only in faster progress of key developments but also results in the training of building labour skilled in modern construction techniques. By setting the initial standards of housing low it is possible to increase the rate at which sound housing can be supplied. The standards set are usually low in the sense that the dwellings are adequate as a shelter but lack the engineering services and finishes usual in Western countries. These can usually be added later when most people have been provided with adequate housing. It is the work on the engineering services and finishes which requires the skills short in developing countries, so that such countries are in a better position to make rapid progress with policies aimed at the provision of basic housing than in attempting to build immediately to Western standards. In the country areas, where the population is less likely to be employed as full-time paid labour, there is sometimes scope for the householder to make some contribution to the construction of his dwelling. Usually some professional help is necessary. This might take the form of professional builders carrying out site preparation and setting out, or perhaps building a frame, the householder being left to clad it. Another method sometimes followed is to integrate the householder labour with the professional labour. Elaborate systems of integration are clearly easier to organise where the householder is seasonally unemployed, or after a widespread catastrophe.

Generally there is little shortage of labour in developing countries, the labour problem being rather a shortage of skilled and organised labour. The value of self-help schemes lie in tapping labour which would otherwise not be utilised and in providing a formal organisation which can increase the supply of housing and other construction and bring it into relation with demand without the need to wait for the full development of a market economy.

Since traditional building tends to require a great deal of skilled labour and since in developing countries this tends to be expensive either absolutely, or in relation to its efficiency, it may be more economic to use some form of prefabricated building rather than *in situ* building. This will depend on to what extent it is possible to use untrained labour to carry out simple repetitive tasks, in prefabricated construction as compared with traditional construction, on the comparative prices of the different materials and on transport costs. In some countries, for example, South Africa, there has been considerable success in dividing the skilled operations into simple, repetitive operations which can be undertaken by labour after very little training.

As the economy develops, the price of labour will tend to rise to levels usual in developed countries. At the same time its productivity will tend to rise so that the price of a comparable unit of work will not rise proportionately. The incomes of the consumer will also, of course, rise so that supply and demand will tend to be in balance at a higher level. As the standard of living rises labour will tend to be more expensive and it will be of increasing importance to use it as productively as possible. This will tend to increase the importance of forms of construction economic in labour as compared with those economic in manufactured materials. The development of building materials' industries and of transport will tend to reduce the costs of manufactured materials and hence reduce their costs in relation to labour costs. As development proceeds, the forms of construction traditional to the area will tend to change, in the same way as in Western countries, towards a professional form of building in durable manufactured materials in which labour economy is of importance. The earlier traditional forms based on abundant labour and cheap local materials will tend to disappear, gradually, especially in and near the cities.

CHAPTER SEVEN

INNOVATION IN BUILDING CONSTRUCTION

INNOVATIONS AND BUILDING ECONOMICS

The importance of the economics of building in relation to the forms of construction cannot easily be overstressed. The relationship between economics and innovation in building construction can be examined most easily by looking at the relative prices of different materials and at the relationship between the prices of labour and the prices of materials. Not merely have forms of construction based on cheap local materials and abundant cheap labour been replaced, as labour costs rose, by forms more economic in labour, but methods and modes of building expensive in labour have tended to fall into disuse. For example, decoration has become far less ornate because of the labour necessary and its increase in price. The carving of timbers and stone, and decorative plaster work have largely disappeared and decorative effects are obtained by using contrasting materials and paint.

In Britain the brick gradually became the dominant material for structural purposes, replacing the materials indigenous to other parts of the country. For example, Scotland has largely ceased to use its native stone, formerly its principal building material, except for aesthetic purposes where additional costs can be justified, and has imported bricks from England and sometimes from the Continent. Construction in stone, even in its cheaper forms, is twice as expensive as in brick.(14) Concrete blocks are often even cheaper for walling, since they can more easily be made in larger units than clay products and hence require less labour for their construction. But their appearance is usually considered poor compared with brick and when they are used externally they are usually rendered; when rendered they tend to be more expensive than brick.(15) In some countries the clay industries have gone some way to producing clay blocks in larger units, which usually results in a lowering of the costs of walling. In this field aesthetics play an important part. Generally, large units of block work are considered unattractive whether of clay or concrete and are usually rendered when used externally. The cost of rendering usually tips the balance of cost and makes the use of block work uneconomic except where rendering is necessary in order to make the walling weather-tight. The use of block work for external walls, therefore, depends on the production of some new form of finish which would be acceptable unrendered. The alternative is some form of block which is sufficiently cheap to bear the cost of rendering; for this purpose concrete may have greater potentials than clay.

Brick and block work has also been largely used on the Continent and tends to be the form of construction used in developing countries during the transition from the traditional form of development formerly indigenous in such countries. Elsewhere, for example, North America and Australia, timber has been an important building material. This usually occurs where there is an abundance of accessible and easily worked timber, where the climatic conditions enable timber construction to provide a durable building, and where either suitable clay for bricks, or the fuel necessary for burning them, are in short supply. For example, in the U.S.A. bricks were found to be four times as expensive as timber, as compared with Britain.[(16)] Again, in Australia bricks are expensive as compared with timber.[(17)] Price ratios change over time, of course.

The prices of labour vary from one country to another as much as the prices of materials. As a result forms of construction economic in one country are not necessarily economic elsewhere. The importance of the effect of labour and material cost ratios can be made clear by comparing the way they have affected the forms of construction in different countries. In America labour was found generally to be more expensive in relation to the cost of materials than in Britain.[(16)] Moreover, skilled site labour has been found to be about 50 per cent more expensive than unskilled site labour and about 5 per cent more expensive than off-site labour working on a corresponding product. In Britain skilled labour is only about 10 to 15 per cent more expensive than unskilled labour and about the same or a little less than the cost of off-site labour producing the same product. Overall, labour has been found to be about twice as expensive in America as in Britain, whereas materials, compared on the same basis, were a little cheaper than in Britain. Generally, America appeared to have a price advantage in timber and structural steel but not in cement while bricks were much more expensive in America.

International comparisons are difficult to make and the interpretation of the results uncertain. It is difficult to obtain data for comparable buildings. Comparisons are often based on small samples of unmatched buildings when it is difficult to differentiate between the large variations between sites and contracts within a country and variations between countries. A review of comparative studies of construction in America and Britain, carried out over a long period, found considerable variations which could not be explained by the available parameters of design and organisation.[(17)] Moreover, comparable prices change and international exchange rates do not necessarily reflect comparative internal purchasing power.

The difference in the ratios of labour and material prices obviously has a profound effect on the relative economy of various forms of construction in the two countries. Traditional timber structures and claddings are more likely to be economic in the U.S.A. than in Britain. Timber lends itself to factory prefabrication, which was found to be particularly worthwhile in the U.S.A. because of the relative low cost of factory labour as compared with the corresponding site labour. Structural steel similarly cheap in itself in America also lends itself to off-site prefabrication. It also was found to be generally used in America for building frames in preference to the use of structural concrete, largely used for this purpose in Britain and Europe.[11] Even so, *in situ* concrete was found to have a comparative advantage in America as compared with Britain because of the skilled-unskilled labour ratio.[16] Similarly, masonry cladding was found to be largely replaced in America by curtain walling which is light in weight and can be largely prefabricated off-site. It is of interest to note the way building regulations can interact with economic factors—the difficulty of satisfying fire regulations with curtain walling encouraged the use of extensive areas of glazing which was exempt from fire endurance standards.[11] Had the running costs of glazing, the costs of cleaning and of replacing the heat losses been considered the conclusions might have been different. Another result stemming from the comparative cheapness of off-site labour in America was the large scale use of ready-mixed concrete.[11] While light-weight precast concrete was used in America for floors and roofs, it is not used on the same scale as in Europe for walls.[11] This is possibly because of the even greater comparative advantage of steel frames and curtain walling.

America is not, of course, an isolated example, for the interplay between labour and material ratios can be traced in the same way for other countries. For example, Australia has been found to have many of the same features as America although the price ratios as compared with Britain appeared to be less extreme. Again with labour comparatively expensive and skilled labour expensive as compared with unskilled labour and off-site labour the expected preference for prefabrication was found. Thus both the forms of building traditional to a country and, of course, the innovations which are introduced are determined to a large extent by the relative prices of different types of labour and material.

INNOVATIONS AND THE PRODUCERS OF BUILDING MATERIALS

Traditionally, the builder took his materials from the ground or from where they grew and fashioned them himself into construction materials. Gradually, however, the preparation of the materials became the work of the specialist and crafts developed to win the materials from their raw state and to prepare them for the builder. Sand and gravel is excavated, cleaned and graded, timber is felled, cut and seasoned, and stone is quarried, cut and dressed. The growth of specialisation tends to reduce the cost of materials, since the specialists have a greater opportunity to become skilled and efficient than the builder and are able to make a greater use of the specialised equipment needed.

At first, the preparation of the materials by the supplier tends to stop at a fairly early stage, the final fashioning being left to the builder himself. However, once a range of requirements becomes standardised and the preparation can be transferred off the site, it is possible for the materials producers to take the production of materials a stage further and to produce what are, in effect, building components. The extent to which this is possible depends on a number of factors. Clearly, the manufacture of building components, prefabrication, is stimulated in relation to the degree to which replication is possible and by the possibility of using special tools and machines for large scale production. Diseconomies tend to arise if the resulting product is more difficult to transport to the site than the raw materials from which it is made; for instance, it may happen that the component has to have added strength in order to stand the stresses of the journey to the site, or is much more bulky.

Clearly, timber is one of the easiest materials to prefabricate and one which possesses most of the requirements mentioned above. It is not, therefore, suprising to find that with the introduction of efficient transport and mechanical woodworking machinery the manufacture of joinery became a factory industry. Over the last few decades the use of manufactured joinery, window and door frames, doors, cupboard fronts, and other items, have largely replaced site-built joinery for most types of buildings. In many countries the prefabrication of joinery has gone further and, for example, timbers are supplied already cut to size for floors and roofs, and roof and floor frames are supplied already prefabricated.

The success of prefabricated joinery has been followed by the use of steel and aluminium for prefabricating many of the items formerly made of timber. These materials were never an *in situ* material for these purposes, its use in this way being dependent on the acceptance of standard sized units. There is now a large sale of steel and aluminium window frames, door frames, cupboards, partitions and curtain walling. Plastic (PVC) is used for window frames in some countries.

Prefabrication has also developed in other fields, particularly for plaster and concrete products. The use of wet plaster is declining, its place being taken by plasterboard and plaster panels. Gypsum is also used as the basis for casting load-bearing reinforced internal walls, which, for example, in Australia, have been used in conjunction with brick external walls.[19] Concrete is used more and more extensively in a precast form, taking the place not only of *in situ* concrete but also replacing to some extent brick and steel. Precast concrete in the form of blocks forms a good substitute for bricks for many purposes. In Britain the concrete roofing tile has largely replaced the clay tile. Precast concrete is extensively used for the manufacture of columns and beams, for flooring, posts and fencing, and for many other purposes. As its finish is improved the need for the addition of another finish will become unnecessary and it will be able to compete even more strongly against the more traditional materials for such purposes. Ready-mixed concrete is another example of an innovation which has progressed very rapidly and it is now extensively used in many countries. Ready-mixed mortar is also now extensively used.

Prefabrication is also important in the provision of engineering services. Plumbing units are often available complete with all their fittings, including pipe connections, as are hot water heating systems. Electrical wiring kits are also prefabricated for domestic buildings.

Not all the innovations in building materials have taken the form of prefabrication. Many have been improvements in the design of traditional materials, or in the form of substitute materials for the ones formerly used. As already mentioned, concrete products have replaced clay products to an important extent. The clay industries have perhaps not developed the use of clay as much as might be expected. Asbestos cement products have tended to replace iron products, for example, asbestos cement sheets have largely replaced galvanised iron sheets for wall and roof cladding.

Manufacturers naturally look for additional markets for their products. The market for building materials is so large that even a small share is very attractive, and so manufacturers naturally look to see in what way their products might be used in the building industry. In this way the plastics industry has found a large market for its products. Because of relative falls in prices and later a much slower relative rise, plastics have become a very competitive material. In Britain, Europe and America it has secured a large share of the floor surfacing market, is providing wall and cupboard top surfacing materials, pipes and rainwater goods, materials for damp-proof courses and electrical and other fittings. Plastics are used extensively as a covering material during the building process. At present its use is limited for structural purposes by its cost in relation to other building materials but as indicated earlier it is now becoming competitive for window frames.

Clearly, the use of new materials is hazardous since their future behaviour cannot be foretold. Laboratory tests can never fully replace the experience of natural ageing and new materials must to some extent be taken on trust. Manufacturers are sometimes prepared to guarantee unestablished materials for the length of life necessary for them to break even in cost with their rivals. In this way their use on a large scale is promoted earlier than would normally occur and the economies of scale can be achieved.

It has been concluded by one research worker that innovation in building materials appears to originate more from a desire by manufacturers to extend or hold markets than in answer to the expressed needs of the building industry.[20]

INNOVATION IN MARKETING MATERIALS AND COMPONENTS

Manufacturers and distributors have changed their marketing methods in a number of ways both in their relation to contractors and consumers.

Packaging is an aspect of marketing. The growth in labour costs and the availability of handling devices has encouraged manufacturers to supply their materials in convenient packs. Bricks, tiles and other units are frequently strapped together and put in plastic bags. This tends to reduce the costs of handling, breakages, damage by the weather, pilfering and other site losses. Packaging kits of units and fixings ready for assembly, for example, for joinery, plumbing and electrical work, reduces site time for erection and from delays arising in the absence of parts and fixings.

The supply of smaller packs, suitable for the consumer market, has assisted and encouraged the home handyman.

As indicated earlier, standardisation and rationalisation of building has encouraged producers to supply prefabricated units. Manufacturers market complete components such as roof trusses, staircases, kitchen and bathroom units both for contractors and the home handyman. Producers also extend their activities vertically to the fixing of their products. Tile manufacturers have contracted for roofing for many years. In a similar way precast concrete and other manufacturers contract to fix their products.

INNOVATION AND THE BUILDERS

The influence of builders on the development of methods of construction depends on their degree of freedom in determining the materials and methods of building. This freedom is severely limited when they are building to the detailed design and specification of a professional designer. However, an important part of building activity is carried out by contractors responsible both for the design and construction. This occurs either where the builder errects off-the-peg buildings, or designs and builds bespoke buildings. Often the contractor has some freedom in deciding on materials and forms of construction even when erecting to the design of a professional designer, particularly with the American type of contract since materials are not always specified closely and substitutes are often acceptable. Within the field of building the contractor probably has most scope in the housing field, and perhaps in the field of civil engineering where many contracts are obtained on the basis of design and construction.

The development of new materials and new ways of using old materials is usually actuated by a shortage of traditional materials or of specialised labour, or by the hope that the newer alternatives will be directly or indirectly cheaper. In Britain the brick replaced such local material as cob, pisé and clay lump, not because it was cheaper as a material, but because far less labour was needed to build with it. Again, concrete block has replaced brick for some uses partly because of brick shortages and partly because of a saving in labour. Prefabricated components have come into use where their overall cost was lower than for *in situ* work. A shortage of materials in the traditional form or of the labour for carrying out the work stimulates the use of new methods. A shortage of plasterers has made the acceptance of plaster board and other plaster products far easier than would otherwise have been the case. Sometimes prefabricated materials have been favoured because they materially assist in shortening the period needed to carry out construction. Early completion brings forward the date of payment to the contractor and shortens the period between the owner laying out his capital and obtaining an income. The use of prefabricated components often make it possible to simplify the whole programme of construction and to dovetail the operations more closely. Contractors sometimes prefabricate on the site to this end.

The extension of the range of materials in the way described above is not usually very rapid. As new materials are produced they are gradually tried out and if satisfactory they gradually become accepted as a part of the traditional method of building. Generally, of course, innovations in the form of new and prefabricated materials are more acceptable the less radically they affect the techniques of building. Innovations in materials necessitating radically new techniques and organisation only usually obtain rapid acceptance when there is a serious shortage of materials or labour, or where some organisation is attempting to find a radically new approach.

The most dramatic change in building methods have probably occurred in those forms of building in which the contractor produces a standard building for sale off-the-peg. This has occurred particularly in the housing field and to some extent for schools and for shed-type factory buildings, where the builder either builds speculatively for sale, or contracts with private or public bodies for the supply and erection of standard building units. Non-traditional methods have been developed under these conditions in many countries, particularly since 1945. This form of development will be considered in the next chapter.

INNOVATION AND CONSUMERS

The growth in the scale and scope of the home handyman as a builder has tended to increase as the cost of contracting building work to professional builders has risen relative to other prices and as desired standards of comfort, convenience and appearance of dwelling have increased and been required by a wider range of housholders. As noted earlier (Chapter Three), prices of building work have risen 50 per cent faster than general prices. The prices of maintenance and improvement work have risen even faster than for new work because they tend to be labour intensive. The prices of such work have been further increased where VAT and sales taxes have been introduced or increased. Manufacturers have greatly increased the advertising of building components and house interiors, considerably adding to the publics' acceptance of higher housing standards and raising the accepted level of expenditure on home improvements. The ability of householders to carry out such work themselves has been increased through such innovations as premixed, non-drip paints, strong and convenient adhesives, easily worked and jointed sheet and pipe materials, kits of parts, prefabricated components and low-priced or easily hired power tools. Increasing numbers of large-scale, comprehensive, retail stores and small specialists now cater for the home handyman. In Britain this market is now put at £1.5 billion a year, approaching an annual expenditure per household of £100.[(22)]

STANDARDISATION AND MODULAR CO-ORDINATION

Clearly, the standardisation of building materials and components will carry with it certain advantages. These will be enjoyed by all the parties to the building process, clients, designers, constructors and manufacturers.

Clients will gain from standardisation both in so far as it reduces costs and in the convenience of making replacements. The difficulty of repairs and construction where no one manufacturer's product will fit with that of another is well known. Often parts cannot be obtained to fit earlier models and a whole component must be scrapped.

The acceptance of a limited range of standard components inevitably has some effect on the freedom of design. Theoretically, the designer is most free when there is no limit to the dimensions he may choose. Much of this freedom is in practice illusory, since where the number of solutions is very large a solution is impossible unless some arbitrary restraints are accepted. In practice, units with particular sets of dimensions can often only be obtained from one particular manufacturer and real choice could well be as large with a rationalised and generally accepted set of standard dimensions.

The acceptance of standardised dimensions is both encouraged and made necessary by the growing practice of using prefabricated units. This has, perhaps, developed furthest in the field of joinery, where it is now common practice to purchase prefabricated window-frames, doors, cupboard fronts and staircases. Clearly, there are considerable advantages in ordering, in designing and in fixing if they all conform to standard dimensions

It seems obvious, other things being equal, that costs per unit will fall the longer the run for any particular article, that is from economies of scale . How large such savings might be will depend on the extent to which rationalisation increases the length of runs. Figures put forward by various countries indicate that costs fall at a reducing rate as the length of the run is increased. For example, Finland quoted experience that in the manufacture of doors costs fell by 14 per cent when a run was increased from 20 to 50 but only by 5 per cent for an increase of from 200 to 500.[21] Another set of figures showed savings of 18 per cent when the run of sanitary goods was increased from 10 to 100 and 6 per cent for an increase of from 100 to 400.[21] The reductions to be expected from such rationalisation in terms of production costs may not therefore be as large as is sometimes supposed and these could be lost if long runs lead to more stocking and an increase in distances for delivery. But many prefabricated units are purpose-made and the runs are very short. In such cases the savings from standard units could be very large. Moreover, standardisation could lead to a reduction in stocks both at the factory and during distribution and this could produce significant reductions in costs.

The logical step following the standardisation of components is to adopt a system of standardisation which relates the standards of all components to a unified set of dimensions. This is much more difficult than standardisation itself since, if the full advantages are to be obtained, agreement must be reached not merely within the same trade but for all trades and possibly on an international scale. At its simplest the idea of modular co-ordination is to plan in terms of a common dimension large enough to facilitate the fitting together of materials and components without waste and, if possible, without cutting, or packing to adjust levels. Modular co-ordination should result in a saving of both labour and materials, a reduction in the range of sizes and should hence enable stocks to be reduced, and enable manufacturers to obtain longer runs of a reduced range of sizes, and hence at lower costs. A further reduction in the range of sizes should be achievable over and above that possible with standardised components. Probably the complete realisation of these ideas is impracticable but significant savings should be possible. There are, however, a number of serious difficulties to be surmounted.

Probably the most serious difficulty is the transition to modular dimensions by the various manufacturers. This could involve large-scale and expensive re-tooling. The greater the use which can be made of current dimensions in the modular system the more readily manufacturers are likely to accept it. The cost of the changeover must be taken account of in considering the most suitable modular number pattern and the basic unit from which it stems. There is a large amount of overseas trade in some materials and components, and some comprise between the inch and the meteric system is necessary.

Difficulties also arise from the lack of dimensional accuracy in traditional building components and in building construction. Many traditional building materials, for example, clay, concrete and plaster products, are subject to shrinkage and warping during production and a high degree of dimensional accuracy cannot be guaranteed. Associated with the dimensional inaccuracies of the materials and components, and partly because of them, are the dimensional inaccuracies usual on the structures themselves; tolerance ranges of as much as 3 in. have been found in the dimensions of rooms in small houses

Such tolerances, while easily accommodated by traditional building techniques, tend to make the use of modular co-ordination futile, since fitting tends to necessitate cutting and waste.

Improving the accuracy of the dimensions usually adds to the costs. The acceptability of tolerances varies with the nature of the operation.[24]

If units are to have a wide range of use, standardisation of jointing and performance standards is also necessary. Units will only have a wide range of acceptability if it is possible to connect one unit to another without difficulty and without unsightliness. Units need to be provided with some form of jointing which is equally workable for a wide range of materials. Similarly, the units to be joined and the joint itself need to possess corresponding performance standards. For example, an external wall unit must provide an equal standard of weather resistance and thermal insulation to the unit to which it is joined and these standards must be maintained through the joint itself. Similarly, for example, internal partition units must provide common standards of insulation against noise and there must be no break in the sound barrier where they are joined.

Experiments have been made in building with modular materials both in America and in Europe, and satisfactory results have been claimed. For example, an American experiment in building houses showed a saving in manhours of 21 per cent, with a cost saving of 10 per cent, but the experiment was rather limited.[23] Other experiments which have been quoted were not carried out simply in terms of modular as against conventional planning, and their effects were confounded with the effects of other changes, so that the results are not conclusive.

Nevertheless, the value of the principles of modular co-ordination must not be rejected because of an absence of experimental proof. Common sense suggests that a soundly thought-out system should result in some reduction in costs. In some countries official backing has been given to modular co-ordination. For example, in Britain a module of 4 in.:10 cm has been accepted with the inch as the smallest preferred increment. For official building in Britain a standard floor to floor height of 8 ft 4 in. with a floor to ceiling distance of 7 ft 6 in. to 7 ft 10 in. has been accepted for housing. The acceptance of such a standard is of particular importance for the prefabrication of components for flats. It makes it posible to produce standard staircases, lifts, dust chutes and wall units which can be incorporated in a large range of designs based on this standard.

The existence of two standards of measurement in the Western world has increased the difficulties of standardisation and co-ordination on an international basis. The trend is to metric measurements. Changing from imperial to metric raises considerable problems during the period of transition, machines have to be retooled. The incompatability of many products produced on different basis of measurement reduces effective choices and creates particular difficulties for the maintenance of such products. Most workers have a concept of the meaning of a unit of measurement with which they are experienced; obtaining an equally sound appreciation of a new unit is one of the problems of transition.

CHAPTER EIGHT

THE INDUSTRIALISATION OF BUILDING

THE RATIONALISATION OF BUILDING

The last two chapters have been concerned with the nature of traditional building in various types of economies, with the economic conditions which have determined the way they have developed in answer to the physical needs of the communities, and the way innovation has gradually changed the forms of building regarded as traditional. While the forms of traditional building tend to change over time, the underlying characteristics tend to remain the same. Traditional building usually implies a loose form of organisation in which the operations follow a recognised order and in which both the order and the work in each operation is well understood by the skilled craftsmen and is implied by the design without any need for detailed instructions. This situation is not necessarily changed in any fundamental way by the introduction of prefabricated components, or by the standardisation of their measurements.

The term non-traditional does not really imply anything about the form of building other than that it is different from traditional. Non-traditional has been used in the sense of prefabrication of buildings and of system building. The industrialisation of building implies something wider than non-traditional. It is concerned with the rationalisation of building, both the types of construction and the building process itself being embraced by the term industrialisation. It is concerned with systems of construction which can be organised in an integrated way, the whole process of construction being organised as a single interlocking process in which materials and labour are available in an organised flow, so that construction can proceed without hindrance as a continuous operation. Industrialisation does not necessarily imply a completely new system of construction. In fact, logically, industrialisation starts with an examination of the traditional form of building and with an attempt to reduce the costs and increase the speed of building so that buildings are ready for use in as short a time as possible, commensurable with cost. This involves organising the constructional work so as to reduce delays and breaks in the continuity of work. This may be achieved partly by the use of mechanical plant, partly by replacing *in situ* work by prefabricated units, and partly by rationalising the organisation. In the process the design may need to be modified. The amount of departure from traditional forms of construction can be so considerable that the design ceases to bear much relation to traditional forms of construction. Industrialisation can take the form of new methods of construction employing a different range of materials, perhaps depending heavily on prefabricated units. Again, it can take the form of a system of building depending on the use of special components or plant only available from the organisation which developed it, and designed perhaps to produce only a limited range of buildings. But, whatever the system or form of building, the process depends on research, experiment and development.

Clearly, industrialisation of building is an aspect of design, of construction and the organisation of the building industry itself. Perhaps rationalisation is a better word to describe the application of science to building than is the term industrialisation.

INDUSTRIALISED BUILDING

Industrialised building implies any form of rationalised system of building in which the main components are designed or selected to form part of a complete system of building and in which design and production are linked, as in the production of most manufactured products. An industrialised system need not be non-traditional; it can be a rationalised system based on traditional methods in which the components are all related and erection follows a predetermined plan. The term could apply to the erection of buildings of bricks, blocks and other traditional materials for which the design had been prepared in relation to the economy and continuity of the structural process. For example, this might take the form of a simplification of the brick and block work to reduce cutting, the use of precast concrete for lintels and steps, so as to avoid a break in continuity, the use of a prefabricated roof structure, so as to minimise the time between the completion of the walls and roofing the building, and the use of precut piping for the water system and for electric conduits. The important principle is not the use of prefabrication but its use in order to obtain a rationalised system giving the maximum advantage right through the construction process. For example, it might not be worthwhile to use precast concrete if this necessitated a lifting device which could not otherwise be used. The use of prefabricated components would be linked to the reduction of breaks in the continuity of other operations and in continuous work for any lifting device. Each design component needs to be examined in relation to the others and in relation to the labour and plant needed, and to the costs. The prefabricated components might be site- or factory-produced; which solution was used would depend not just on the cost of production of the component but also on the way in which the production linked with the rest of the construction process. For example, site production might be used even if this resulted in the component being more expensive, if, as a result, overall savings were obtained perhaps by using the production of the component to use labour which would otherwise be idle. The emphasis should be on the reduction of overall costs which generally requires continuity, a reduction in the number of separate operations, and high utilisation of labour and machines; every feature of the design and production process needs to be examined in relation to the whole process. The logic of the design-production process may lead to considerable departures from traditional forms of building, perhaps to the use of factory-made components in quite new materials, perhaps to an all-dry form of construction, perhaps even to pre-assembled rooms, or complete housing units which need only to be connected to the services on arriving on the site.

It is, however, clear that the more industrialised building results in a large-scale use of special capital equipment, the more it is only likely to be economic either if a large number of similar units of buildings are required, or if a large building is required which lends itself to the repetition of similar units. Housing, and to some extent schools and small industrial and storage buildings, tend to lend themselves to large-scale production, and large industrial buildings and tall buildings with identical floors can be based on repeated units. Clearly, unless there is a large amount of repetition it is unlikely to pay to carry out the detailed research and development necessary to rationalise a particular system of construction. Moreover, this rationalisation involves integration of design and production with the implication either that the designer must work from the outset with the contractor who is to erect the building, or that the designer and constructor must be within a single organisation. The first solution is hardly compatible with competitive tendering, and the second implies a limitation on the freedom of the client in his choice of design. In fact, the client would normally have to choose between a limited number of available buildings—off-the-peg buildings; only if his demand was on a large scale could he expect the contractor to produce buildings with special features. The extent to which special buildings could be produced would depend on the flexibility of the system of construction; the less it depended on large-scale prefabrication or on special moulds and formwork, and the more it was based on materials and components available on the market, the more likely it would be that special design requirements could be met.

PREFABRICATED AND OTHER NON-TRADITIONAL BUILDINGS

The prefabrication of buildings is by no means a new idea. Perhaps it started with the use of iron as a structural material. Certainly there are records of the prefabrication of buildings going back a great many years, for example, the lock-keeper's house at Tipton, England, built of iron in the eighteenth century; the iron palace of King Eyambo in Africa and the iron departmental store in New York were both built in the nineteenth century.[25] In the second half of the nineteenth century there were such notable examples as the Crystal Palace in England, the prefabricated barracks used by the British in the Crimea and subsequently re-erected at Aldershot, and the package township sent out from England to Australia.[25] In the 1860's several American firms are reported to have supplied prefabricated buildings ready for erection.[32] Large panel systems similar to those used today have been in use in Europe and America for the last 70 years.[26] But the development of prefabricated buildings did not reach large proportions until the 1920's.

Between 1919 and 1944 some 52,000 non-traditional dwellings were built in Britain, but this was only about one per cent of the dwellings which were erected in that period and only about half of them were prefabricated, the balance being of *in situ* concrete.[27] During the early twenties the use of non-traditional forms of building was given official encouragement, although no special financial rewards were offered. It was hoped that non-traditional forms of construction would supplement the traditional forms of construction and so increase the output of the building industry. The emphasis was on systems which would use little brick and perhaps more important, ones which would make little use of skilled building labour, then in short supply, but which would make use of factory labour which was in plentiful supply. It was hoped that the buildings would be no more expensive than traditional buildings. In the main, four types of systems were used, based respectively on *in situ* concrete for walls and floors, on timber and metal frames, with timber and metal cladding, on special purpose made concrete building blocks and on large factory made units.[27]

It is difficult to determine how successful these systems of building were, for few records of their performance and costs appear to be available. Their inability to survive the period of greatest housing shortage suggests that in general they did not win the acceptance of either the public housing authorities or the public. Sir Lawrence Weaver,[28] writing in 1926, considered that they did not offer any real advantages over the traditional house. Freemantle,[25] considered that the ultimate costs of the non-traditional houses, when account had been taken of their probable life and maintenance costs, were unlikely to compare favourably with those of well built traditional houses. The same conclusion was reached by two government research workers in 1933.[30] In 1942 an Interdepartmental Committee[27] was set up by the Government to examine methods of house-building in relation to post-war needs and the conditions likely to prevail. In the course of their study the Committee examined the pre-war experience. Broadly, all the types reviewed reached a certain level of acceptance and passed the building by-laws. But they contained faults, such as a liability to excessive moisture penetration, excessive heat loss, and inadequate sound insulation. Some of these troubles were regarded by the Committee as teething troubles, which could be cured by small changes in design and specification without altering the principles of the design. The merit of the methods reviewed, of particular importance in the expected post-war situation, was that the methods were economic in skilled building labour, the burden of the labour being transferred either to unskilled site labour or to off-site labour. Claims that the alternative methods provided a house that could be erected in a shorter time could not be substantiated in the absence of data. The price data available was not very conclusive because at the time that the houses had been erected there has been little competition. Many of the houses were finished in roughcast and were similar in appearance to traditional houses.

The systems differed from traditional construction only in so far as the structure itself was concerned; the finishes were normally similar and were usually applied in a traditional manner. Of the 26 systems examined by the Committee all used their own special system of construction for the external walls and over a half of them used it for the party walls. More than half of the systems used traditional construction for partitions and chimney breasts. Few of the systems had other than traditional first floors and roofs. Apart from the timber houses few systems avoided some use of wet plaster, either internally or externally, as roughcast. With certain exceptions joinery was traditional as were the services.

In general, not more than a half of the construction differed from traditional, and in most cases the proportion was only a quarter. The smallness of the element of non-traditional construction was a virtue as well as a limitation. While it reduced the possible saving in cost over a traditional house it also limited the chances of the price getting much above it. In the main, the prices of the non-traditional houses were about the same or a little above those of traditional houses; one system using pig iron was priced out of the market when the prices of its main materials rose. The Committee thought that maintenance costs were generally similar to, or a little higher than, for traditional houses.

Again, after the war, in 1945 Britain turned to non-traditional solutions for house construction. The Committee[27, 31, 32] examined the systems put before it and approved no less than 101 systems as suitable for development or for use by local authorities. These systems, in general, followed the lines of the pre-war systems, some of which were included in the list. The three main types were again concrete, steel-framed and timber houses; 46 were based on precast or *in situ* concrete, 35 had steel frames and 13 were timber-framed or solid timber houses. Unfortunately, in the main, only prototypes were examined and no estimate of costs under normal production were published.

A large number of schemes were approved by the Government and special capital grants were allowed to meet some of the expenses of tooling and converting factories for the prefabrication of the components.[33] The grants varied from £50 to nearly £250 per dwelling. The grant for the aluminium bungalow, a wholly prefabricated building, was £708 per dwelling.[34] This was equivalent to about two-thirds of the cost of a traditional dwelling. The policy of allowing these grants proved very expensive and they were withdrawn at the end of 1947.[33] After that schemes were approved only if they could compete in price with traditional buildings, as, in fact, some types had from the beginning. From the end of the war up to 1955 about a fifth of the houses built by public authorities in England and Wales were non-traditional; in Scotland the proportion was about twice as high.[35] For a time after 1955 non-traditional dwellings still in use were regarded as being equally acceptable as traditional. But it is believed that the proportion of prefabricated buildings declined considerably. One estimate puts the proportion of non-traditional to traditional dwellings at about 6 per cent for England and Wales in the period 1961-2.[36]

Non-traditional dwellings came into favour again in the mid-sixties when the government again sponsored system building. At the height of their popularity about two-fifths of public authority dwellings were system built. The proportion was highest for tall multi-storey flatted blocks, for which no traditional form of building existed and least for two-storey houses. The prices of system built dwellings, on a comparable basis, were generally a little above those of traditional dwellings, but this may reflect the market situation rather than costs. In the early seventies their use started to decline as multi-storey flatted blocks, and system building itself, again fell out of favour. The use of systems declined further as overall demand for dwellings declined. As demand declined their prices became more competitive, but despite this few system built dwellings were used by the mid-1970's

It is not known whether non-traditional dwellings have been used for private enterprise building, although it is known that building societies tend to be reluctant to advance money on building other than those of traditional construction. Private purchasers also tend to be biased in favour of the known type of housing. Of course, the forms of construction accepted as traditional tend to widen with time

Speculative house-builders have used various forms of rationalised construction to improve productivity and reduce site time and costs. The market conditions of the early nineteen-eighties have created additional difficulties for speculative development. Traditional construction tends to tie-up large amounts of money in partly constructed dwellings over long periods; money which is often difficult and expensive to borrow. By the time dwellings are ready for sale the market may be less buoyant and prices may have fallen. Builders can reduce both their overall costs and market uncertainties by reducing site time and preferably building to order. It is claimed that the use of timber frames reduces erection time to about 6 weeks, so that orders can be executed while the purchase process is proceeding.[37] A substantial increase in the use of timber frames is anticipated.

In non-housing fields, particularly factory construction, steel and concrete framed buildings now represent the traditional form of construction.

In addition to the non-traditional dwellings built for permanent use, a number of types of bungalow were developed for temporary use. These can be divided into two classes, package bungalows and the aluminium bungalow. The first class consisted of about ten types all of which were based on factory-made, prefabricated, light-framed sections of metal or timber. Externally, the frames were covered with metal or asbestos cement panels, with plaster board and fibre board internally, with a backing of aluminium foil or some other material for thermal insulation purposes. The internal partitions also consisted of plasterboard or fibreboard on frames. The floors again were timber-framed as were the roofs and these were covered with asbestos cement sheets, metal or roofing felt. This type of bungalow was taken to the sites in sections and erected on previously prepared bases, after which the fixings and fittings were added and the dwellings were given their final decorations. The aluminium bungalow was quite distinct from the other types and was the most highly prefabricated. It left the factory in four component units, each of which was nearly complete with the fixtures and fittings in position. Whereas the framed sections of the other types were first stockpiled to await distribution to the sites, the sections of the aluminium bungalows were conveyed direct from production in the factory to the site, and were unloaded and directly positioned on to prepared foundations. The four sections were then coupled up, cover boards and a few other loose sections were fixed, the services connected and the final decorations were applied. The actual erection time was only about 2 hr, while the entire site erection work took only about one and a half manweeks.

Most of the temporary bungalows were a little lower in cost than the permanent houses being erected at the same time, but had floor areas only two-thirds as great. In 1947 the estimated average cost was around £1170[38] for a temporary bungalow as against £1250[33] for a permanent traditional house with land, roads and services; the figure for the aluminium bungalow was around £1600.[38, 39] But the temporary bungalows had a much shorter potential life, although, no doubt, with careful maintenance their lives could be considerably extended. On the other hand, their price allowed for refrigerator, washboiler, immersion heater, cooker and rather more elaborate cupboard fittings than usually provided in permanent houses. There is little doubt that as the programme worked out it was a very expensive experiment. On average the bungalows in the first class cost around £355 at the factory, site works and erection cost £332, transport, distribution and contingencies cost £208 and fittings £275. The aluminium bungalow cost around £1236 to produce at the factory, site works £240, transport £68, and erection £53.[40]

Clearly, the aluminium bungalow was far too expensive to compete with the traditional house even if it has been as durable—in fact, in many of them the main members corroded badly within about 15 years and the dwellings had to be removed. The other types showed more promise but even so it is difficult to argue that they provide a potentially competitive permanent house. Perhaps £170 of the cost could be discounted as a measure of the value of extra appliances over the traditional house. On this basis there would have been available about £250 to convert the structure to one with a normal life and with a floor area a third greater. While, in terms of prices at that time, it would have been possible to replace the outer cladding with a brick skin and the roof with a traditional one of timber and tiles for the sum, there would have been little likelihood of being able to increase the floor area within the same cost margin.

On the Continent the characteristic form of housing is the flat, not the house as in Britain. For example, in France the ratio of flats to houses is more or less the reverse of what it is in Britain. In Continental cities a large proportion of the dwellings are provided in the form of five- or six-storey flatted blocks with walk-up access. Hence on the Continent it was the flatted block on which attempts at industrialisation were centred, whereas in Britain it was the house. Again, on the Continent the principal difficulty was the shortage of skilled labour, and industrialised methods were developed, as in Britain, to supplement the output produced by traditional means. In France, while there was a shortage of skilled labour, cheap unskilled labour from Italy and Algeria was available.(41) The shortage of skilled labour was also a basic reason for turning to industrialised methods in the Scandinavian countries and in eastern Europe. In these countries there was the additional incentive provided by the length and severity of the winter which seriously interfered with production by traditional methods. In Western Germany there was no shortage of labour until about 1958, additional labour being freely available from southern Europe and from eastern Germany. While labour remained plentiful, wage rates remained moderate and German builders concentrated on the rationalisation of traditional methods, particularly increasing the level of productivity in brick and block work. Where there was no shortage of labour, as in Italy and Spain, little interest has been shown in industrialisation.

Shiffer(42) estimated that in France, Denmark, Sweden and Holland somewhere between a tenth and a twentieth of present housing construction was by industrialised methods, mainly in the form of multi-storey slab blocks with a box frame of large precast concrete panels. This is, of course, a considerably lower ratio than attained by non-traditional dwellings in Britain in the period 1945-55. In Germany little use has been made of industrialised methods until the sixties. On the other hand, in eastern Europe industrialised methods have been used on a large scale. It is thought that about a half of the housing constructed in Moscow and Leningrad was built by industrial methods and that by the mid sixties half the housing in the whole country was built by these methods.[43] However, this is easier to achieve under the conditions pertaining in eastern Europe where the level of uniformity in the design of housing is far greater than would be accepted elsewhere. Often thousands of identical blocks are erected on the same site.[44] In both Britain and America the popularity of the single family house sets a limit to the comparative importance of industrialised methods of flat construction. In Britain, about 10 per cent of local authority houses, 45 per cent of local authority flats, and about 15 per cent of schools were being constructed on the basis of system building in 1964. System building is only important in the public building sector.[45] In America such methods are applied to the individual house and it is claimed that a fifth of these houses are wholly prefabricated and that prefabricated components such as floors, wall panels and roof trusses are used in half the other housing.[46] But, of course, houses in America, as on the Continent, tend to be relatively expensive as compared with traditional houses in Britain.

SYSTEM BUILDING FOR FLATTED BLOCKS

The systems used for the construction of flatted blocks fall into a number of classes. The number of different systems is very large, running into some hundreds, and to list them would serve no useful purpose. Of the concrete systems perhaps the large panel type of system is the most well known. These have been usually based on factory-made room-sized panels. In systems like the Coignet, the panels and other precast concrete units are made with great precision and fit together easily. In other systems, such as the Camus system, the units are less expensive to make but are less accurate and more time is usually necessary to fit the units together on the site.[42] Both systems were widely used in various parts of Euorpe and there is some development of this type of industrialised housing in South America.[42] Usually the panels are cast in the form of a sandwich construction incorporating insulation material; service pipes and conduits are also cast into the panels. Special trucks are needed to convey the units to the site and a crane is required for their erection.

Another range of systems have been based on site-cast concrete panels, for example, the Swedish system, Sundh. Usually the site-cast units are less precisely finished than the factory-cast units. The casting on site needs to be much simpler and is subject to the hazards of the weather but it is not necessary to maintain special fleets of transport to the sites and handling and transport costs are lower. The site-casting plant is usually cheaper than the factory-casting plant but is less likely to yield a high level of productivity. Most of these large panel systems use heavy load-bearing cross walls as their basis, although the Swedish Fastighets system uses beam and column and light-weight panels.[42]

Systems have also been based on combined *in situ* and precast concrete elements; the French Sectra system is an example. In this system precision-made steel formwork is used to place structural concrete in rectilinear tunnel sections of room height and width.[42] The formwork is heated so that it can be struck and re-used in about 13 hr. The finish to the concrete is said to be adequate for direct decoration. It is claimed that a floor of five flats can be completed in 2 days. Some constructors have used a formwork which is kept intact and having been erected for the first of the floors is then made to climb up the building from floor to floor and is not struck until the top floor has been poured.

There are also systems for flat construction based on steel frames, for example, the German Hoesch system, which was developed for low blocks of flats. This system is based on prefabricated light-gauge steel sections with precast concrete floor slabs, light-weight concrete slabs for walls and plaster board partition units. By using largely standard materials the heavy investment of capital for a precasting plant is avoided[42] In Holland a system, H.S.S.B., is used based on steel reinforced brickwork.[42]

Attempts have also been made to prefabricate complete room units. The Swedish Skanska Cementgjuteriet has developed the heart system consisting of bathroom, W.C., boiler room and part of the kitchen around which the rest of the dwelling can be assembled.[47] Heart units have also been developed in the U.S.S.R. and in Britain. The U.S.S.R. has also developed prefabricated housing units of several complete rooms which arrive at the site completely assembled and need only to be lifted off their transport on to a prepared base to be built up into blocks of flats.[48]

OTHER TECHNIQUES FOR FLAT CONSTRUCTION

There is a great variety of other techniques, some of which are not closely related to any particular design, although, of course, they do have some design limitations. One technique is based on vertical casting on site. The first pair of panels of each type are used to build up a battery of panels, which are subsequently used as the moulds for casting the panels needed to build the block; this battery of panels is finally built into the construction.[49] Another technique, the British Jackblock, consists of *in situ* concrete reinforced floors which are successively poured on the ground-floor slab and then jacked up by means of hydraulic jacks to form the floor above; precast concrete blocks are inserted to form the core walls. The floors are cantilevered out about 15 ft so that the external walls do not carry any loads, and block or curtain walling can be used. In this way the construction of each floor takes place on the ground from the top floor downwards, and as the higher floors are constructed work can start on their fixings and finishes.[50] Jacking has also been used with lift slabs constructed on the plate system, in which the floors are cast as plates on the ground and jacked in position up *in situ* columns.[51]

SYSTEM BUILDING FOR HOUSES

Naturally, system building has been applied to the numerically most important type of housing, flats on the Continent and houses in Britain. Some of the systems developed in the earlier post-war period have continued to be used, perhaps in a modified form. They have become familiar and have been absorbed into the accepted range of building methods. The new interest in system building in Britain in the earlier sixties, and the support given by the Government, led to the development of further systems for building houses. Many of these have not been greatly different from systems developed earlier. However, the industrial environment is now very different from that pertaining when the earlier systems were developed; materials such as timber and steel are no longer in short supply, new materials such as plastic have been developed and exploited for building and there are no longer wartime industries seeking new fields of activity, although some shipyards are interested in developments in this direction. In other countries, for example, America, where timber is cheap and plentiful, development has taken the form of standardised and prefabricated timber houses.

Whereas a large proportion of the systems developed for multi-storey construction have been based on heavy units, many light forms of construction have been developed for houses. Often steel frames have been used as a basis, sometimes having wall units of timber-framed panels consisting of a sandwich construction and covered externally with timber or a plastic-coated material. Framed houses are, of course, flexible and admit of many internal arrangements. Timber stud frames have also been used for constructing load-bearing walls and these have sometimes been rendered externally on the basis of expanded metal. Use has also been made of cross walls of precast concrete. In Sweden use has been made of prefabricated brick panels. Aluminium has also been used for the prefabrication of houses, for example, in Australia for use in the tropics. The range of systems which has been suggested is very wide, although many have not progressed beyond the prototype stage.

SYSTEMS FOR OTHER TYPES OF BUILDING

Some of the systems developed on the Continent for housing have been applied to other building types. For example, the Cauvet system has been used for housing, offices, flatted factories, hospitals and large stores; this system is based on a very flexible system of moulds.[52] Marley Concrete, a British firm, has marketed two precast concrete systems suitable for small halls, lecture rooms, offices, industrial buildings and other business and social uses.[53] Another system based on space frames and prefabricated panels has been used in Britain for constructing barracks and has been said to be suitable for many types of buildings.[54] In the U.S.A. prefabricated timber buildings have been developed.[55]

Undoubtedly, in Britain the most important field for industrialised systems outside the housing field has been the schools programme. Probably the first in the field was the Hertfordshire County Council who developed a partly prefabricated system based on a steel frame with brick walls. The idea was taken up by the Ministry of Education whose development group, in 1949, produced steel frame and concrete frame systems with various types of cladding; generally the systems were developed commercially by the firm who built the prototype. The Nottinghamshire County Council, faced with the problems of mining subsidence, as well as with the usual shortages, developed the ideas further with a hinged steel frame capable of overcoming the problems resulting from mining subsidence. The buildings were designed to a planning grid as were other systems for schools, and a complete system of components was designed. Although the whole school programme for the county was committed to the system, the orders proved to be too small to secure the full economies of production and other authorities were invited to add their orders to those going to the manufacturers of the components. Eventually, in 1958, a consortium was formed of a number of educational authorities who agreed to use the system and to combine their orders; the list of authorities was not confined to authorities in areas subject to mining subsidence. The consortium, CLASP, has gradually increased in size with the result that the orders have increased enabling economies of scale to be obtained.

There are about a dozen systems of industrialised building available for school construction. All have to conform to the national cost limits set by the Ministry of Education. However, in spite of the interest shown in system building for schools only about a seventh of the school building consists of system-built schools; about half the schools are still built in load-bearing masonry and the balance are mainly framed with masonry infilling.

SITE WORKS AND PUBLIC UTILITY SERVICES

Less attention appears to have been given to the rationalisation of site works and public utility services than to the industrialisation of the buildings themselves. Some studies have been made of various types of foundations and of the best ways of laying out such services as electric power. Within the curtilage, foundations, public utility services and other site works need to be planned as a whole. They all result in disturbance to the site and tend to increase the difficulties of operating over the site and to interfere both with the erection of the buildings and with their own installation. It would appear to be just as necessary to co-ordinate plans for the services outside the buildings as within them. In some countries, thought is now being given to the design of ducts suitable for a range of services. A common duct should make it possible to reduce the amount of excavation and disturbance to the site and might result in an economic system of accessible ducts and so eliminate future disturbance to the ground works. Full benefit could, perhaps, only be obtained if the linking of services was carried outside the curtilage to the estate distribution networks. The co-ordination of service requirements would probably be stimulated if their installation became the work of a single specialist type of contractor. Perhaps, in some cases, all ground works should be carried out by a specialist engineering contractor, so that their siting and programming could be fully co-ordinated.

CHAPTER NINE

THE ECONOMICS OF INDUSTRIALISED SYSTEMS

THE PROBLEMS OF ECONOMIC ASSESSMENT

The economic comparison of different types of building is not easy. The only type of information generally available is price data and this is not usually very conclusive for two reasons. Firstly, prices vary considerably because of the variability of design, site and market conditions, and unless the prices to be compared differ by a considerable margin it is not possible to be confident that the prices of one type are likely to be consistently less than those of another type. Secondly, unless those responsible for the purchase of different systems of building believe one type to be inferior to the other types, they will tend to be prepared to offer the same prices for both types, so that unless there is a strong incentive from the producers of the newer types to undercut the established types they will all tend to sell for the same price, and any differences in costs will be obscured. Where a new type of building is as acceptable as an existing type and the supply is small compared with the total market there will usually be little incentive to try to undercut. The price of the traditional product will tend to set the price of the new product, and under these conditions the producer of the new product will have to accept a lower margin if his costs are higher, but will be able to retain any savings over the cost of the traditional product. The benefit of lower prices will only tend to be passed on if competition becomes keen either between the traditional and non-traditional products, or between the producers of new products trying to capture the market. Such a situation might arise in a number of ways, for example, if there were a series of technical failures with the newer methods, if the output of the traditional part of the industry could meet demand in quantity, quality and at the right speed without the assistance of other methods, or if strong competition developed between firms using the other methods. In Britain, and in most other countries, most system building has been developed by particular construction firms, and hence each system tends to be available only from one producer, or from other firms building under licence. System building does not necessarily imply proprietary building but it is likely to develop in this way. The battery system developed by the Building Research Station in Britain[49] is not protected by patents and is free for general use. It is, however, likely that most systems will be proprietary since the development of a system is expensive and the developer wishes to obtain a return for the resources he has used. The position would be much the same if development was in the hands of professional consultants. Inevitably, with proprietary systems contracts are usually negotiated and hence tend to be less competitive than for traditional dwellings, the contracts for which are normally put out to tender.

In fact, as will have been appreciated from the last chapter, non-traditional methods have always been introduced at a time when the demand exceeded the supply which the traditional industry could provide. In the past the use of such methods, mainly system building, died out when the traditional industry was once again in the position to meet the demand. It is, therefore, relevant to ask whether at this time the use of industrialised systems will be permanently established. Price data offers very little assistance.

As would be expected, the prices of system-built buildings are usually about the same as those of traditional buildings, the differences probably being generally within the scatter of prices normally found in traditional building. Some of the prices quoted tend to be misleading since they relate to the building without foundations, site services and land, whereas the prices for the traditional product usually include these things.

In the 1920's the system-built houses ceased to be produced in Britain when conditions returned to normal and it was possible to build all the houses required traditionally and at normal prices. Again, in the 1940's system buildings came in to help meet the shortage at a time when traditional building was still disorganised after the war and costs were high. As the traditional industry recovered the system-built houses tended to become more traditional. By 1955 the system-built houses were much the same in price as traditional houses, although they were a little smaller. There is no evidence either from the Continent or from Britain that the prices of the system-built houses or flats are any less than those for traditionally built dwellings. Writing in 1964, the National Economic Development Council concluded that the main attributes of system building were that they economised in scarce skilled labour and that they shortened the process of building.[55]

Reporting in 1964, the chief architect to the Ministry of Housing and Local Government for England and Wales said that industrialised building systems find it extremely difficult to compete in terms of cost with traditional building; at best they are comparable with the costs of traditional housing.

But despite the limited success in developing systems of building which compare favourably with traditional building in cost, it would be foolish to dismiss system building as of no interest. The development of system building has had its effect on traditional building in providing a challenge, in stimulating new ideas and in developing ideas which have been absorbed into traditional building. Moreover, because system building has been developed only in periods when the traditional building industry was disorganised or unable to meet needs for other reasons, it has had to operate under difficult conditions, as well as enjoying the opportunity such times have offered. Often the systems developed appear to have arisen to meet transitional conditions instead of the normal ones. For example, firms have developed systems to utilise labour, plant and manufacturing knowledge made redundant by the end of orders of armaments: systems have been developed to meet needs when traditional materials were scarce. Again, many forms of production have failed several times before success has been obtained, often because the conditions for success were at first absent.

EXPERIMENTAL APPRAISAL

Unfortunately, little attempt has been made to carry out scientific studies of the economics of system building. Clearly, it is difficult to carry out experiments in this field since, if adequate replications are to be obtained, a large volume of building must be carried out. It is far cheaper to obtain statistical information by observing normal production but difficult to obtain an adequately balanced sample.

Studies carried out in the mid-1940's indicated that the systems then in use for house construction in Britain saved skilled site labour, but the evidence for a saving in costs was less clear-cut.(52) Of the systems considered only two based on *in situ* concrete showed any saving. While a number of systems based on precast concrete units were only marginally more expensive, this was at a time when the prices of materials and labour used in the traditional houses were relatively high.

Later, the Building Research Station carried out a series of planned experiments with three systems of construction which incorporated what were thought to be the most promising ideas in the field.(58) None of them proved as cheap as the traditional dwellings which were built under the same conditions. The results of the experiment did, however, suggest some important factors necessary for the success of systems of construction. The materials used in prefabricated unit are often a good deal more expensive than those used in traditional *in situ* construction. Labour costs must fall sufficiently to offset the extra costs of materials if overall costs are to be no higher. Since, even in traditional building, materials are about twice as expensive as labour, considerable economies in labour are needed to offset any rise in the cost of materials. The trend is for the prices of manufactured goods to rise less rapidly than the price of labour, so that the use of prefabrication and components which require less labour to fix should become more economic over time. In fact, in some cases the prefabricated units were both more expensive than the traditional materials they replaced and actually cost more to erect. The main reason why prefabricated panels failed to reduce costs was because they only replaced straightforward parts of the traditional work; the complicated work had still to be carried out traditionally. As a result the number of operations was increased as was the time taken in stopping and starting. The introduction of factory-made components only appears likely to be economic if whole traditional operations are thereby eliminated. For example, the use of panels for partitions and the inner skins of walls did not eliminate the use of wet plaster around chimneys and openings, and the use of blocks left the need for external rendering. Again, since factory production does not appear to be necessarily substantially cheaper than site production, very little extra can be provided for the raw materials and the rationalisation of the factory production, and its scale needs considerable attention. Moreover, some contractors appeared to be successful with systems which differed but little from systems which appeared unsuccessful in the hands of other contractors. This suggests that organisation is a key factor in the success of system building.

TECHNICAL AND DESIGN FACTORS

In the face of the historical evidence of system building over the last six decades, it is perhaps difficult to understand the extent of development along these lines and the confidence which has been expressed in system building in some quarters. There are a number of explanations for this situation. The most important, perhaps, has been the scale of the demand to be met.

As a result of the war many buildings were destroyed and the normal building programmes were suspended for 6 or 7 years, greatly aggravating both the shortage of buildings and the quality of the stock available. Further household fission has increased the number of households to be housed per unit of population. The pressure to rebuild the urban environments was, and to a large extent still is, considerable. Demand was further increased by rapid increases in the population. Hence any methods which promised to enable the rate of building to be increased were very welcome and received strong political support, particularly if costs were not increased. Moreover, many laymen find it difficult to accept that there are not more economic methods of construction that those based on such small units as bricks and blocks. For these reasons, and because of the natural appeal of ingenious methods, many people are strongly predisposed to any new system of building.

Perhaps in some ways the approach of sponsors and designers has been too radical. Perhaps too much reliance has been placed upon technical ingenuity and too little regard has been paid to the experiences which led to the evolution of the traditional house. The engineering solution of mass production cannot easily be applied to something as large, heavy and intricate as a building, which is also highly individual and cheap in relation to size. Buildings do not easily lend themselves to flow production—generally the work does not flow to the operative, he must move round the work. Furthermore, the basic traditional materials, bricks, tiles, sand and aggregates, and even cement and plaster, are low priced products in relation to their bulk. Their price is low because the labour content for winning them and converting them is low. But the dimensional accuracy of such products is also low and some of them are subject to shrinking and warping. The difficulties caused by low dimensional accuracy can be overcome fairly easily when building *in situ*, since the materials can be built into each other. Any unsightliness can be covered by renderings and plaster work and by the use of cover mouldings. Cover mouldings provide a way of covering shrinkages while at the same time allowing the material freedom to move, so that further shrinkage and warping do not cause ugly cracks. But it is these finishing trades which are so expensive, particularly in labour.

In general, those attempting to rationalise the design of buildings have taken as their aim the reduction of labour requirements, in particular the reduction of skilled site labour. The accepted solution has been to increase the size of the building units, to eliminate the finishing trades and to prefabricate as far as possible off the site.

The use of large prefabricated units demands dimensional accuracy and materials which remain free from distortion. Such materials tend to be expensive, either because, like metal and plastics, they are highly processed materials, or because, although employing cheap basic materials, for example, aggregates and cement, careful control is needed in order to obtain dimensionally accurate components. A change to dimensionally accurate materials is more likely to increase the cost of materials than to reduce the cost of labour. If, in order to obtain materials suitable for prefabrication, more expensive materials are used, it is likely that a prefabricated unit will cost a good deal more than the traditional materials it replaces, especially when the costs of prefabrication are included. An increase in the cost of the materials by a sixth, roughly needs a reduction of a third in the costs of labour, in order for the cost of the new method to break even with the cost of the traditional method.

The scope for savings in changing from traditional to prefabricated units does not appear to be large. The most promising approach appears to lie in trying to eliminate the finishing trades and thus to reduce the number of operations. This is often possible where the units are dimensionally accurate since the joints can be exposed. It is not usually possible if the system is comprised partly of prefabricated units and partly of traditional construction. Hence the attraction of systems which keep traditional and prefabricated construction separate either by building entirely in dimensionally accurate units, or perhaps by using unit construction internally around which a separate skin in traditional work is placed.

ECONOMIC FACTORS

Overheads per unit of labour are generally greater in a factory than on a site because the annual costs for buildings, heating and lighting and for plant and welfare are usually much greater per man for labour employed in the factory than on the site. Against this, output per man is usually greater in the factory because more technical aids are available, organisation is easier and the weather does not impede production. But both output and overheads per unit of output depend on the scale and length of production run for each product. For the production of buildings the problem inevitably arises of securing orders sufficiently large to secure much economy in overheads.

The market for buildings, even for public housing, tends to be too diverse to provide orders large enough to secure the full economies of scale, although the size of orders which would be required is not very large. Housing programmes are developed on an annual basis by large numbers of separate clients. Generally each site is considered separately with both individuality and variety having importance. Clients are free to place their orders with any acceptable producer and usually require product modification to suit their design preferences. As a result orders tend to be small, erratic and uncertain. Rationalisation of the market would require long-term ordering of standardised buildings from a limited number of suppliers in each regional market. The economies, which such an arrangement should enable to be realised, might not be passed on to the purchaser if competition was too limited.

The relationship of capital to other costs is of importance in determining the economics of the different types of sytems. It has been suggested that the amortised cost of special equipment is only of the order of about 3 per cent of the costs of each unit.(59) The relative proportion depends, of course, on the frequency of use, but it does appear from the few figures available that the amount of special capital required for system building is not very great, certainly in relation to other industries. As a result, increases in output beyond a moderate number will tend to make only a little difference to the costs. One investigation of a factory prefabricated system indicated that most possible savings in capital costs had been obtained with an output of as few dwellings as 800 by which time the capital costs were about £57 per dwelling, a doubling of output to 1600 dwellings reduced the costs to about £40, and a further doubling to about £30.(60) The numbers of dwellings given are the total number, not the number per year. Thus a large production of each type of system may not be important; the economies of scale may, in fact, be rather small. If this is so the danger of monotony from the repetition of similar units in the same part of the country may not be as serious as it is sometimes suggested.

Of course, an increased use of moulds can be obtained without the repetition of similar buildings. This can be achieved in a number of ways. The simplest is to change the external treatment incorporated in the wall units. Another method is to use moulds which can be adjusted to handle a range of sizes and shapes, although this is likely to add to their costs. A third possibility is to produce smaller units which can be used in a large range of designs; the traditional brick and block is the ultimate in this direction. Clearly, as the block is made smaller the costs of handling and site erection per unit area will tend to rise and the advantages from large-scale production will tend to be lost. It may be that the most economic solution lies somewhere between large units developed for a system and traditional units manufactured for stock and used by contractors for a range of designs. Clearly some system of large units flexible enough for general use, what is called an open system, would restore to the designer some of the freedom he loses with closed sytems of building.

One of the most fruitful ways of reducing costs is to eliminate operations, for example, by incorporating the services into the wall and floor units, or by producing units which are self-finished. This is difficult, if not impossible, in the case of a general contruction unit; impossible for services since continuity around the block cannot be obtained, and difficult for finishes because of the number of separate units in each wall, floor or ceiling. Technically, joints present one of the greatest difficulties; the greater the accuracy and the better the fit the less the time necessary in erection and in making good, but the greater the cost of production. Up to a point, traditional prefabrication provides some of the advantages of the open system. Once a few design limitations, such as a standard floor-to-floor height, have been accepted it is possible to extend the range of standard units. The Scandinavians, who accepted this at an early stage, produce for stock standard precast staircases, refuse chutes, wall units, floor and ceiling units, and bathroom and kitchen units.[(61)]

Transport costs can add considerably to the delivered price of factory-made building units, particularly if the units are heavy or weak. The additional transport costs have to be set against the advantages of concentrating output at one place.[(62)] While some wastage of materials is eliminated as a result of factory production, the weight carried to the site may not be less, for example, large quantities of water are absorbed in concrete and plaster goods. Furthermore, materials reaching the site via the factory may travel further than those going direct. Light-weight sections may not be cheaper to transport than heavy ones because, with less rigidity, greater care in packing and supporting during transport, and in subsequent handling is necessary. For one proprietary system, Reema, a typical example of a system based on precast concrete units, the cost of transport per dwelling for a journey of about 100 miles is given as £100, about 5 per cent of the costs of dwelling.[(63)] It has been suggested that even for a journey of 10 miles the transport costs per dwelling might be £30 and that for a journey of 25-30 miles they might well be double.[64] The cost depends less on the mileage than on the number of journeys per day each transport vehicle is capable of making.

Factory production is not, of course, essential, for the units can be prefabricated on the site, or *in situ* methods can be used and thus the transport cost eliminated. There remain the costs of transporting the materials to the site but this may be no more expensive than transporting them to the factory. Site production may not be as efficient as factory production but the cost of the site plant is generally less than the cost of a factory and its plant. Some methods based on site production use very little special plant. For example, the battery method of site casting requires only one set of moulds for each type of unit, since the battery for producing the units of construction consists of a set of building units themselves, the units finally being built into the construction.[49] Some systems based on *in situ* construction also use the same set of moulds repeatedly; in fact, such systems differ from traditional *in situ* concrete construction only in the standardisation of the moulds to allow repeated use.

Undoubtedly, part of the economy of industrialised building as it has developed as system building has been obtained from the logic and orderliness of the methods of production and erection. This in itself may contain a lesson for traditional building. Some contractors believe that the success of the industrialised systems is more a factor of the organisation than of the method of fabrication.[65] Certainly some construction systems, for example, "no-fines" concrete, were more successful in the hands of some contractors than others.[57]

Traditional methods appear to be more flexible and can more easily meet fluctuations in demand than industrial system building. Long contracts are required for system building because of the way costs rise if factories operate at a low level of output—a high proportion of factory costs are fixed, since often neither the labour force nor the overheads can be reduced easily when output falls below the optimum. Cost might, therefore be reduced if sufficient orders could be obtained to allow each system to work at its optimum level. The organisation of consortia and the grouping of orders should help to provide orders of the optimum size but perhaps only if orders are restricted to a few systems. Clearly, large sites can only be obtained in the development of new or greatly expanded towns or in redeveloping large clearance areas, and it may be that this is the field in which systems building should be concentrated. An order covering several sites is equally advantgeous for factory production but not for site erection.

It is often argued that if the contracts for traditional building provided for runs of similar units of building as long as those for system building there would be no difficulty in devising systems of organisation which would enable both delivery dates and costs to be greatly reduced.

While factory production reduces the hazards of the weather, these hazards can also be reduced on the site by the use of polythene tents for the protection of the operatives and the work, and by the use of additives to cement and mortar and heated formwork, although such devices have their own costs.

There is some speculation as to whether the departure of industrialised system building from traditional methods has gone far enough to secure really worthwhile economies. Generally, the industrialised systems are based on the same range of materials as traditional methods. Savings in material costs are, therefore, marginal. Most systems are based on concrete. As described earlier, materials such as metals and plastics are much more expensive unless used in very thin sheets. But it is difficult to obtain the required rigidity, or thermal and sound insulation from thin sheets. Moreover, there is no certainty that such materials possess the durability and weathering properties required. While light-weight and precision materials might enable a reduction to be obtained in handling, transport and erection costs, these, it will be remembered, are the lesser part of the costs of buildings. This situation is in contrast to that in some manufacturing industries where, prior to the introduction of mechanisation, the costs of labour formed the major part of the costs.

As mentioned earlier, one of the aspects of industrialised methods is the substitution of one type of labour for another. The two principal types of substitution are unskilled site labour for skilled site labour and factory labour for site labour. While such substitutions may enable additional types of labour to be obtained, they may do little to reduce costs: this will depend on the relation between the gross cost and output of labour of different kinds. If there is a large differential between the wages of skilled and unskilled site labour, as there is, for example, in the U.S.A., where skilled labour tends to be paid about 50 per cent more than unskilled labour, or in France, where the extra is about a third, then a substitution of unskilled for skilled will be worthwhile unless the output of the unskilled is appreciably lower than for the skilled. In Britain, in contrast, the differential is only about 10 to 15 per cent. Again there are possibilities of substitution of factory labour for site labour, either because grade for grade factory labour is cheaper than site labour, or because unskilled or female labour can be used in the factory, whereas skilled labour would be necessary on the site. Again favourable conditions for such substitutions do not exist in Great Britain. Such economies will, of course, be affected by the differences in labour oncosts.

A change in labour and material rates could then increase the comparative cost advantages of methods based on factory production. Again, new products based on non-traditional materials might prove more economic in industrialised system building than in traditional building. Apart from such changes there appears some doubt as to how far industrialised system building will prove cheaper than traditional building. The possibility of tapping new fields of labour with industrialised system building may not be a unique advantage. It may be possible to rationalise traditional building, that is, to industrialise it, so that new sources of labour could be used. Such a development could be a result of the division and specialisation of labour so that labour could be trained rapidly to carry out simple, standardised and limited operations. It is on these lines that other industries have approached the substitution of unskilled for skilled labour and reduced the time required to manufacture their products.

MANPOWER, COSTS, PRICES AND PROFITS

The available data for comparing the labour content of traditional and non-traditional building is limited. Often labour expenditure is reported only for single experimental sites; the experiments are not repeated and there is little data available to examine how representative such sites are and how much variation there is between sites. For Great Britain the most comprehensive analysis covers 138 sites with about 14,000 dwellings.[(66)] The comparison suggests that for 1 to 4 storey blocks of housing site mandays were about 20 per cent lower for rationalised traditional than for traditional, with little difference between rationalised traditional and industrialised (Table 9.1). Broadly dwellings were defined as traditional if they had load bearing external walls of brick, as rationalised traditional if they had crosswalls of load bearing brick or block with infill panels in the external walls and non-traditional if they were system built.

The main savings in labour for dwellings in blocks up to 4 storeys as between traditional, rationalised and industrialised were for brick-laying with 33, 16 and 10 mandays respectively and labourer with 54, 46 and 46 mandays. There were smaller savings for services and finishes. It is notable that there was little difference in the times for plant operators, about four mandays in each case.

For high flats of five storeys or more there was proportionally less difference between traditional and nontraditional construction (Table 9.1).

The absence of data on prices and costs has already been commented upon.

Such price data as is available is often rather misleading since it applies to the building unit and excludes the services and often also the foundations. There is generally little evidence either for Britain or for the Continent that the prices of industrialised system buildings are any lower than for traditional buildings. On this most writers are agreed. There is also general agreement that, as buildings, they function no better and often not as well as traditionally constructed buildings. Their attraction lies, in fact, in the possibility of using them to increase the total output of the industry. It is necessary to consider why, under these conditions, so many contractors have attempted to produce systems of building.

TABLE 9.1

SITE MANHOURS PER DWELLING

		Traditional	Rationalised	Industrialised
1 to 4 storeys -	Operatives	177.2	136.1	133.3
	All site			
	Labour	187.5	143.8	141.2
	percentages	100	77	75
5 and more storeys				
-	Operative	237.4	-	199.5
	All site			
	Labour	247.7	-	205.5
	percentages	100	-	83

SOURCE: Lemessany J. and Clapp M. A., *Resource Inputs into Construction-Building Research Establishment* - CP 76/78.

Profit is measured against capital employed, and hence a comparative small saving in costs, if retained by the contractor, could represent a satisfactory return on the capital employed. In a paper analysis of a system of construction, which was based on figures from a number of systems, it was found that a saving of 3.5 per cent on the costs of traditional methods was possible; this was equivalent to a return on the additional capital used of 30 per cent.[60] A saving of this order would be a strong attraction to a contractor. While this figure was derived from a paper exercise, variation analysis suggested that the figure would not be greatly affected by changes in the assumptions.

To the contractor a successfully developed system offers a higher return on his capital, a method of construction normally less vulnerable to the weather than traditional systems, perhaps greater certainty in his estimate of costs, and a proprietary product with the advantages which this normally brings in terms of a more stable market. The spread of the use of industrialised systems is likely to increase the importance of large national contractors and contractors who can provide a complete construction service at the expense of the medium-sized, local contractors. Generally the smaller firms carry out repairs and maintenance work and small alterations and are not likely to be much affected.

Generally the professional side of the building industry is doubtful about the value of industrialised system building. However objective they may be in their doubts about the appearance and quality of the finished product and of the chance of the client obtaining a reasonable price, it cannot be forgotten that the general adoption of industrialised system building of proprietary form would seriously reduce the part they traditionally play in the building industry. The problem of poor design and of monotony is a very real one and must be increased with the inevitable repetition of design unless each was prepared by the best designers. Again, clearly, a produced article is less likely on balance to meet the needs of the users than a purpose-made article, but, of course, particularly in housing, only a small proportion of users have specially designed buildings. As against this loss must be set the gain in national efficiency from the use of standard designs. The force of these objections is much reduced with the open systems of building which provide a kit of parts which can be assembled in many different ways and so allow the designer to operate in much the same way as with traditional building, although without as much freedom. There is no evidence of whether the open systems are any less efficient than the closed systems. A considerable range of design is also possible with those industrialised forms of building, which depend on technique rather than prefabricated units, for example, jackblock, plate systems and the battery systems.

The absence of evidence of industrialised system building being cheaper than traditional building does not necessarily indicate very much about costs to the community. It is, of course, very difficult to establish differences in prices of buildings because of the large amount of variability in prices from site to site. Moreover, it is difficult to be certain to what extent the price reflects costs. Clearly firms can charge up to the market price set by the traditional methods of construction. During the early years of a particular system market prices may not cover the full costs though later costs may be well below market prices. Thus, as long as purchasers do not discriminate against system building such buildings will command the same price as traditional building and their sponsors may obtain very high profits in relation to the capital employed. If this were the situation there might be some additional merit in the use of industrialised systems from the point of view of the community, since there might be some saving in the use of national resources, although there might be objection to the sponsoring firms making above normal profits. It is difficult to evaluate and compare buildings and hence difficult to be sure whether the industrial building offers as good or better value for money compared with the traditional building. There can be no certainty until the buildings have stood the test of time.

It seems to be generally agreed that the design of the system-built buildings makes a fuller use of the properties of the materials than does traditional building design, and hence there is generally less reserve strength to provide for unforeseen contingencies. Many of the systems are very dependent on the connections between the units being carried out exactly as designed, and a failure to follow the design exactly is often more serious than for traditional work. Difficulties are also experienced in ensuring weather-tight joints, which is much more difficult in a damp climate, such as experienced in Britain, as compared with the drier climate experienced on the Continent.

Both systems building and innovations in traditional building use both new materials and existing materials in new ways. Failures have been experienced in both types of building, for example, in trusted roofs and in vertical claddings.

Clearly these new methods have not been in regular use long enough to gauge the durability of the buildings and their component parts. However, the rates at which failures have been reported, the higher costs of early maintenance and repairs necessary to make the buildings safe and the incidence of failures where demoliton is less expensive than repair, suggest that the incidence of failure and future lives will be less favourable for system built dwellings than for traditionally built ones, particularly in the case of high flatted blocks. Especially for high flatted blocks, costs-in-use appear likely to be substantially higher than for traditional houses. They are half as expensive again to build in Britain than houses and maintenance services costs appear to be running much higher. However, their costs to operate and their rates of demolition also reflect their lack of social success. It is for this reason that many high blocks are being demolished.

CONTRACTING ARRANGEMENTS FOR SYSTEM BUILDING

As mentioned earlier, sponsored system bulding tends to put the contractor in a stronger position in relation to the purchaser than is normally the case with traditional building. Whereas contracts for traditional building are normally put out to competitive tender, there is usually only one contractor available for any particular system of construction and thus no competition. Each system still, of course, competes with the other systems and with the traditional forms of construction but it is difficult to compare differences in prices between different designs and differences in sites. The position of the purchasers can be strengthened if they combine. Moreover, if the purchasers combine their orders and organise them into a properly balanced programme the costs of production will tend to be lower for the contractor. Some of these savings are likely to be passed on to a strong group of purchasers. Thus a consortium of purchasers is not only in a position to bring a countervailing power in its negotiations with the contractor, but by organising its purchases into a continuous programme may reduce the costs to the contractor and use the power created by its size to obtain the benefit of such reductions. Again, if its size is adequate it will be worthwhile to set up a techncial group to make a thorough study of the alternative types of buildings available. Often the consortium arrange not just one contract but a series of contracts in which are incorporated rules for fixing the future prices. This is, of course, a way of organising the continuity of work which enables the contractor to spread his capital costs over longer periods and hence to offer price reductions.

System building need not stem from a contractor, although usually it does. Equally well the sponsor can be the client. For example, in Britain one of the best known sytems of construction has been developed by the client, in this case the local authority education committees. As described earlier, the CLASP system was developed with a hinged steel frame to meet the problem of building schools in an area subject to mining subsidence. The system proved to be acceptable even in areas not subject to subsidence, and other educational authorities were invited to join with the sponsoring authority to provide large orders. In this case the system was developed by the purchasing authority. They designed the system, the components and the methods of erection. The components were designed in consultation with the manufacturers to simplify production. The consortium places long-term orders for components matched to the amount of construction and contracts with building contractors to erect the buildings where they are required.

Interpretation of the economic results of this development is difficult, although, at least in this case, prices are available. Over the first 4 years of the scheme the prices per square foot for CLASP buildings fell, while building prices in general were fairly stable. There seems to be no doubt that the CLASP buildings can be provided within the same price ceiling as other types of school.(67) The fact that the prices per square foot were a little lower for CLASP than for schools in general is not conclusive evidence of their being cheaper, since the floor areas were greater for the CLASP schools. The standard of finish of the CLASP schools is said to be at least comparable with those of other schools, whether they will be as durable only time can show. The CLASP development has introduced three changes and it is interesting to speculate on which of these are responsible for the success of CLASP. Firstly, the system substitutes prefabricated, dry finished for normal, wet *in situ* methods. Secondly, the components are purposely designed for the system instead of being purchased to fit a separately designed building. Thirdly, the system provides bulk, long-term orders to the manufacturers and continuous contracts to the contractors. It is difficult to determine which of these elements contributes most to the economy of the system. It is, of course, possible that a traditional system would give equally good results with purpose-designed components and bulk ordering.

SYSTEM BUILDING AND THE FUTURE

Perhaps both the virtues and the disadvantages of the various types of system building have been exaggerated. At present there appears to be no evidence that system building offers any advantage in price, and while system building does reduce the site time there seems to be some doubt whether the overall design, or ordering to completion time is any less than with traditional building. The main advantage lies in the possibility of increasing the national rate of building, during periods when this required by bringing into the industry additional sources of labour without the need for them to serve an appreticeship. This advantage is most marked with the factory systems for which it is comparatively easy to bring labour from outside the construction industry. Where the units are made on the site, construction labour is probably usually used, but even in these cases there is probably some substitution of unskilled for skilled labour. But as long as the systems still use as many resources as traditional methods there may be no overall gain to the community unless the builders use the labour to better purpose than it was used before; this would be true if, for example, the labour was previously unemployed. There might be no overall gain if the labour was merely drawn from other industries. The additional buildings would have been obtained at the expense of less output of some other goods.

The economic advantages offered by factory construction would increase if the prices of site labour rose compared with that of factory labour. Such a rise might occur either because site wage rates rose disproportionately or because of an increase in site labour incentives or improved conditions of working.

It might be just as easy to reorganise the work of the traditional forms of building and to follow the usual industrial process of the division and specialisation of labour, so that the craft processes could be broken down into a number of simple operations, each of which would be easy to learn, and so that no long period of apprenticeship would be necessary. In various ways this is the path of evolution which the traditional building industry has been following. The use of skilled labour has been reduced by the use of prefabricated materials and components, for example, the use of plasterboard and plaster panels for wet plaster, the use of prefabricated joinery and metal fittings for *in situ* built joinery and the use of prefabricated fireplace surrounds for *in situ* tile construction. This process could be taken a great deal further. Moreover, for housing and many other types of buildings the amount of skill needed in a craftsman is declining. Carpenters and joiners no longer fashion elaborate work on the site; modern adhesives and metal fasteners make jointing largely unnecessary; bricklayers are only called upon for the most straightforward types of brickwork and plasterers only carry out straightforward plastering. Mechanical plant and particularly powered hand tools reduce the need for some skills as well as reducing fatigue. Generally it is only in the rare new building and in restoration work that the craftsman can exercise his full range of skills. For most work it is possible that craftsmen of adequate skill could be trained much more rapidly than at present.

While the advantages of the systems appear to be fewer than often claimed, so perhaps are the disadvantages. The problems of monotony can be handled in a number of ways. A great variety of external treatments can be incorporated into the external wall units and a good appearance can be secured if sufficient care is taken, although clearly this is likely to raise the cost. The systems can be adapted to a variety of designs, but the greater the number of designs the more moulds and templates will be needed, and these will tend to raise costs. However, as shown earlier, beyond a certain point costs are not greatly affected by increases in the length of run, and the additonal costs of greater variety have probably been exaggerated. A full utilisation of the factory capacity over its life may be of most importance.

It is necessary to consider how fast even apparently economic innovations should be introduced into construction. Technological studies and experiments can only expose the more obvious risks of failure. Aging tests can never be absolutely certain; some types of failure will only reveal themselves as the building matures. On the other hand, if utilisation is left until there has been sufficient time for materials to be tested over their life cycle, resources will be wasted through continuing to use materials and constructions less satisfactory than the new ones. In the case of system building political priorities have often played an important part in decisions to use them. In Britain, nontraditonal methods, broadly system building, provided about 45 per cent of new local authority dwellings in 1967-8 but the proportion had dropped to about 20 per cent by 1978.[66]

The movement of opinion against system building in Britain arose in part because system building was largely applied to high flats. As a result the building failures appeared and to some extent probably were more dramatic than failures in two storey dwellings. Moreover, public opinion moved against high flats because of social as well as technological failures. In the meantime, many of the more successful innovations have become a part of accepted traditional techniques.

CHAPTER TEN

THE MECHANISATION OF CONSTRUCTION

As EXPLAINED earlier, system building is only one of the ways in which the building industry is being industrialised. Whatever the form of building, productivity can still be increased by rationalising the site processes. This is a problem partly of mechanisation and partly of the organisaiton of the work on the site. The two are related. It is convenient in the first place to consider the type of plant available and the part it can play.

MECHANICAL PLANT

Mechanisation is another form of industrialisation. Greater mechanisation has often been advocated as the simple answer to the problem of increasing productivity in the building and civil engineering industry. Advocates of mechanisation have labelled the industry backward, because on any measure the amount of mechanical power used is low compared with that used in manufacturing industry. It is often overlooked that the building industry is an assembly industry, the component parts being manufactured by other industries. Usually it is easier to use powered plant in manufacturing than in assembly. Generally, the amount of powered plant used on the assembly side is a good deal less than on the manufacturing side, particularly where the products are not very homogeneous, for instance, in the assembly of aircraft and ships. The assembly of buildings is subject to the additional hazards of work on location, to the effect of frequent changes in the location of working, often over wide geographical areas, and to the effect of climate and open-air conditions. Moreover, the nature of the work varies and with it the opportunity of using any particular item of plant.

The proportion of the output of the industry for which mechanical methods are applicable is smaller than often appears at first sight. In many countries nearly a third of the work carried out by the industry is repairs and maintenance. Little powered plant can be used for such work beyond the use of powered hand-tools for paint spraying and such work as drilling, planeing and sawing, and of course, breakers and surfacing machines for road work.

Generally, in this sector mechanical plant plays only a small part in the work. Of new work at least a half consists of finishing trades and services for which, again, only small powered hand-tools have much relevance. The possibilities of using mechanical plant are greatest for the structure of the buildings and for civil engineering work, but even so, the proportion of work for which mechanisation is possible is often small as can be seen by examining the work content for a traditonal small house.

As indicated in the previous chapter the main reductions in the labour content of a dwelling as between nontraditional and traditional construction lies in bricklaying and labouring. Because most dwellings are still largely traditional in their construction and as much as a seventh of the work arises from site works which are not affected by the form of building, the relative importance of the various trades does not change very much. This is indicated by figures for building in Great Britain over the last three decades (Table 10.1).

About two thirds of the work for a traditional house is skilled craft work a quarter of which is carried out by bricklayers and most of the balance by carpenters, plasterers, plumbers and painters (Table 10.1). It is difficult to visualise any mechanical plant which could assist bricklayers; mechanical hand-tools can be used to assist the carpenter, electrician, painter and plumber: such tools reduce the physical work and some of the skills needed. Mechanical appliances have been developed for spraying plaster and renderings, but their use has been delayed in some countries by the difficulty in obtaining satisfactory finishes. Similarly, paint sprays have proved to be of only limited value unless large areas of a common finish are required. The unskilled work consists mainly of general labouring work and bricklayers' labouring, with a small element of plasterers' labouring and other trades. It is because of the large element of skilled craftsman's work in traditional construction, and of the difficulty of mechanising it, that the use of other forms of construction, more amenable to mechanisation, are frequently advocated.

The number of man-hours of work involved in labouring is, of course, very variable; it depends not only on the level of productivity but on the nature of the design, on the site and on the extent to which mechanical plant is used. The use of non-traditional methods has less effect on unskilled and semi-skilled labour expenditure than on skilled and little effect on use of plant operations.[66] Generally, about a third of labouring time is used in excavation, site clearance and levelling; about a half the time is used in handling and moving material, and the balance in mixing materials and general jobs such as cleaning up after an operation is complete. Soil movement is perhaps the easiest operation to mechanise; materials handling and unloading can also be mechanised, as can the mixing of materials. The mechanisation of the work of the trades' labourers is much more difficult since it consists of a large number of different types of jobs, such as handling small quantities of materials over short distances and in confined spaces, and in cleaning up. In all, perhaps only about a third of the work is capable of being mechanised. As a result, in the case of a traditional house only about 10 per cent of the costs can be affected by mechanisation. Since plant itself has a cost it would appear to be necessary to reduce man-hours by perhaps a half to two-thirds in this sector to obtain a net saving of the order of 3-4 per cent in the costs of a house.

TABLE 10.1

TRADE BREAKDOWN OF OPERATIVE TIME FOR A DWELLING

	Percentages of operative time for Local Authority Traditional dwellings			
Sample Periods	1948-51	1954-56	1962-7	1967-73
	Random	Selected	Selected	Random
Trade				
Bricklayer (incl. labourer)	34.3	27.6	27.6	28.0
Roof tiler	2.1	1.5	3.2	1.5
Carpenter	14.7	11.6	13.5	13.9
Plumber & Heating engineer	5.6	6.5	5.8	6.3
Electrician	2.1	3.2	4.1	4.8
Gas fitter	0.7	0.5	0.5	0.5
Painter	9.5	8.0	8.4	-
Plasterer	9.9	12.3	7.7	7.5
Floor layer	0.8	1.0	1.5	0.8
Unskilled, semi-skilled, other	22.2	25.8	27.5	28.0

Figures as quoted, not rounded off.

SOURCE: Lemessany J. and Clapp M.A., *Resource inputs to construction: the labour requirements of house building*. Current Paper 76/78, Building Research Establishment, 1978.

The possibilities of increasing productivity are much greater in some forms of building and with some forms of construction. Clearly, the larger the proportion of the job which is structural the greater the possibilities of using mechanical methods. Again, the greater the element of structure which consists of large units or of poured *in situ* concrete the greater the opportunity for using mechanical methods. The advantages of mechanical methods are not limited to gains from the direct substitution of mechanical for hand operations, but include the possibilities of gains from the integration of the operations. In fact the best results cannot be obtained simply by the substitution of mechanical plant for hand operations: the job must be looked at afresh to see the possibilities of the whole construction process when mechanical aids are substituted for hand methods.

The effective costs of plant in the construction industry tend to be a good deal higher than in manufacturing industry. This arises from a number of causes. Firstly, the plant must be taken to the job rather than the job to the plant. Often the amount of work for which a unit of plant is used on a site is quite small, so that the cost of transporting and handling the plant form a large overhead on the costs of using it. Secondly, while the volume of work carried out on a site may be small, the plant may need to be on the site a long calendar period in order to complete the work, since sometimes there are long intervals between each of the uses of the machine. To some extent greater machine utilisation can be obtained by taking the machine to another job and later returning it, but this is not always worthwhile. Thirdly, job variability both within and between jobs tends to result in low plant utilisation. Finally, the use of contractors plant on rough surfaces and in the open air subjects it to wear and tear far greater than is normally experienced by plant used in factories. Generally, it is used with less care and is less well maintained than plant used in factory work.

The level of plant usage tends to be very low. A study in Britain indicated that plant is not usually out on contracts more than a half to two-thirds of the year the rest of the time it is at the depot either under repair or awaiting repair, or waiting for a site to need it.(68) Even when on a site it may stand idle on more days than it works. Even so, the plant is not necesarily productive for the whole of the recorded working time. Part of the idle time, all that in the depot other than repair time and part of that on the site, is a result of carrying of a wide range of type and size to meet widely varying needs and a fluctuating work load. Idle time also arises from the intermittent demand for plant for most operations.(68)

Nevertheless, in spite of the difficulty of using plant in building, it has an important part to play, because while actually working it can often do as much work as several men, can continue tirelessly hour after hour and can be used to simplify the organisation of the site and to lower costs. It is now generally appreciated that mechanical plant is not just a substitute for labour but a different method of carrying out operations which suits different techniques. Quite a lot of study has been given to methods of using plant and to the development of plant so that it meets site needs more readily.

Again, while the machine may greatly increase the level of productivity of the operation, costs may not be reduced because the organisation carrying out the operation uses the machine so rarely that its effective costs are very high.

This problem can be met either by hiring the machine from a plant hire organisation or by sub-contracting the operation to a specialist who can employ such plant frequently.

The relationship between the cost of the plant and the work it is capable of, in relation to the cost and output of labour, will differ from one country to another and it may be worthwhile using a particular piece of plant in one country but not in another. The higher the earnings of the labour suitable for carrying out the operation manually in relation to the earnings of the labour manufacturing and maintaining the machines, the more worthwhile the substitution of the machine for manual methods is likely to be. Thus, in a country where building labour is paid well, as compared with factory employed engineering labour, for example, America, it will pay to use machines to a greater extent than where building labour is paid less well in relation to factory labour, for example, Britain. Clearly, in developing countries, where for example, the machines have to be imported from high-wage countries but where the local contracting labour is paid low wages, it will not pay to use machines to replace labour to the same extent as in countries both manufacturing and using the machines. In the early stages of development it may pay such countries to use labour in what may appear a very wasteful way. This may, however, produce the best economics results if the effective exchange value of local labour for machines is very low. On the other hand, of course, the value of an early completion date, for example, for a capital work needed to open up an important national resource, intended to provide large-scale employment and output, might justify the use of labour-saving methods even though the cost of the job, looked at in isolation, was much higher as a result.

EARTH-MOVING PLANT

Mechanical plant probably has the greatest potential in earth movement, that is, in digging, excavating, levelling and moving soil. This sort of work requires physical force rather than skill and it is here that the machine has the most to offer. In relation to its costs the earth-moving machine can carry out many times the work of the man it replaces. Its output in relation to cost is so great that it often pays to employ a machine to replace hand labour even when the job is very small. For example, it often pays to bring in an excavator to dig the foundations of even a pair of houses, even though the cost of transporting the machine to the site may be greater than the cost of hiring the machine. Not merely does the earth-moving machine do the work more cheaply than manual labour but the work is done much more quickly than would be possible even with a large labour force, and thus helps to reduce the overall site time. The earth-moving machines are particularly of value in civil engineering, for example, in road building and pipe laying.

The actual output of a machine depends not only on its own potential but also on the limitations imposed by the operations which follow. For example, because of the slower rate of building foundations and the danger of trenches subsiding, foundations cannot be dug in a continuous operation unless special steps can be taken to complete the foundation work in step with excavations. It does not necessarily pay to do this, even though this would be the most efficient method of excavating. While there may be economies of scale for an operation, it may not be worthwhile to realise these because of the diseconomies introduced elsewhere in the building project. The economies of scale may not be an advantage unless the project is reorganised, perhaps by sub-contracting the operation to a specialist.

MATERIALS MIXING PLANT

The main materials to be mixed are concrete, mortar and plaster, of which concrete is the most important. In manufacturing precast concrete and plaster goods the rate of mixing can usually be matched with the output of the mixer and high rates of productivity can usually be obtained. For *in situ* work this is far less easy. The limitations imposed by the following operations are usually greater in the case of connecting operations than for earth moving. The effective output of the mixer is limited to the rate at which the concrete can be poured. For example, the output of a concrete mixer per working day is lower when the concrete is used for pouring columns than when used for pouring floors and walls;[69] some of the highest rates of mixing are obtained when the concrete is used for road works and runways.

Since the speed of mixing per working day is dominated by the rate at which the concrete can be placed, it is difficult to utilise the full capacity of concrete mixers unless the size is closely matched to the rate at which the concrete is required for placing. This can only be achieved if a wide range of sizes of machine are available and this may result in very low machine utilisation, so that, in fact, the maximisation of machine output per unit of operating time may not produce the best results. It may be far better to obtain outputs considerably below the potential of the machines. Again, of course, it may pay to hire machines to match the rate of output required.

Concrete can be purchased ready-mixed. This practice is widely followed in America, Britain, and a number of other countries, and is spreading in other parts of the world. For Britain it is estimated that about two thirds of the concrete used on sites is obtained ready mixed.[70] Whether or not ready-mixed concrete is cheaper than mixing on the site depends on the availability of the raw materials and their location, on the location of the ready-mix concrete plant in relation to the job, on the amount of concrete required, and on where the concrete is to be placed. Generally, it tends to pay to use ready-mixed concrete when it is required intermittently over a long period, providing that the concrete is required in batches equivalent to the capacity of the concrete delivery trucks. But the saving in handling costs in buying ready-mixed concrete often exceeds the saving in mixing costs. This tends to be particularly so where the concrete is required to be placed either at ground level or below, when often the ready-mixed concrete delivery truck can place the concrete direct into position.

With the mixing of mortar and plaster the economy in purchasing ready-mixed is largely obtained from the convenience of ready-mixed material. Of course, in the case of all ready-mixed materials advantages are obtained in terms of the reliability of the mix and perhaps a reduction in the wastage of material.

MATERIALS HANDLING PLANT

Materials need to be handled both horizontally and vertically. Many materials are handled several times and the total weight of materials which need to be handled even for quite a small job is often very high. For example, it has been estimated that the average traditional house contains nearly 150 tons of materials, some of which has to be handled several times.(71) Where little or no powered handling plant is used, as much as 10 per cent of the total manhours required to erect a house may be used in handling materials.(71) Of these materials about 90 per cent have usually been used before the building is roofed. This is fortunate since it is difficult to feed materials through window openings. The traditional methods of handling materials are to use wheelbarrows for horizontal movement and hod-carriers or baskets and jenny wheels for vertical work; methods still used in rural parts of less developed countries. For small jobs, for example, for house construction, it is often difficult to find any cheaper way of handling materials unless the labour is paid at high rates compared with labour in other industries. The use of mechanical handling for small buildings in developed countries is often more because of the difficulty of getting labour to carry materials than because of its cheapness. In developing countries it rarely pays to use mechanical plant for moving materials for small buildings. There are a number of reasons for the efficiency of labour for these purposes. The operative with the hod, to a lesser extent the operative with the wheelbarrow, can take the materials from the stock pile direct to the point of final use. In contrast, double handling is often necessary with mechanical means. When the operative has completed the movement of materials he can be employed on some other operation, whereas the mechanical handling aid is frequently very inflexible and is idle for much of the possible site time.

There is a wide range of horizontal handling plant. At one extreme lies the motorised barrow and at the other full-scale railway systems. The mechanical barrow is a machine for the small job and the small builder. It has only a small capacity but it is cheap to purchase and run, carries considerably more than a hand barrow, and takes the strain from the operator's work. Above this size of machine are the dumpers, some of which have a capacity of many tons; trucks are also used where the surface is suitable. The smallest type of railed transporter is the monorail which is very useful for handling fairly small quantities over a regular route. Above the monorail in size lie the various types of traditional railway systems; these are mainly only used for the largest types of civil engineering works. Conveyor belts are also used and special pumps are used for fluid materials such as concrete, but again only when large quantities are to be moved. There are also special pieces of apparatus such as the concreter boom which was developed for concreting the oversites of pairs of houses. However, as mentioned above, many of the problems of handling concrete horizontally can now be dealt with conveniently by the ready-mixed concrete truck.

The main mechanical aids for the vertical handling of material are hoists, elevators and cranes. The smaller hoists are usually mobile and can be pushed around the site as required; the larger ones are tied into the scaffolding or to the building itself. Elevators are also usually mobile. There are a large variety of cranes, some stationary, some mounted on rails, some on wheels or tracked, and some which are built to climb up the well of a building. In the past most cranes had their jibs fixed near the base of the machine; this improved their stability but limited their usefulness for lifting to a height; the machines had to be placed away from the building and had to be very large for the weight to be lifted. The majority of cranes used in building today are either tower cranes or climbing cranes. Both have booms fixed to the tops of masts; this gives them a long reach and a large range of heights. Stability is obtained by loading the base of the crane with heavy blocks or by tying the crane to the building; where this takes the form of a climbing crane the machine can handle materials for a building of any height. The tower crane was developed in the fifties and its use has spread widely in Britain and on the Continent. The rapid increase in its use is partly because it provides an efficient lifting machine for buildings of several storeys and partly because of the increase in the construction of tall buildings.

The output of any materials-handling machine again depends very much on the rate at which the materials moved can be utilised. This applies particularly to machines handling concrete and machines handling materials direct to the point of assembly. Except where cranes are stock piling they are usually handling either concrete to the point of use or large units. In both cases they become, in effect, a part of the erection process and their output depends almost entirely on the speed of the operation to which they contribute. It was found, for example, that output does not increase with the size of a tower crane, since output is determined by the following operation; the larger machine is used to obtain a larger lifting capacity and a greater reach. The output depends largely on the type of load and the way it is used.[(69)]

Until the introduction of the tower crane, cranes were not used a great deal for building except on the very largest jobs, because they did not provide a very convenient way of vertical handling. The spread of the tower crane has probably had a larger impact on the erection of buildings than most other types of machine. Not only have they made it easier to handle large building units and large sections of formwork, thus facilitating the spread of systems of construction using large units, but they have had a considerable effect on the organisation of work on the site. The tower crane is expensive compared to most other machinery used by builders, as distinct from civil engineers, and contractors felt it important to keep the machine in operation so as to obtain good value for money. This led to the work on the site being planned around the functions which the crane could carry out. If fact, the importance of ensuring that the crane was fully employed has often been over-estimated; the most economic results are not usually obtained by minimising the costs on one piece of plant or on one section of the work. This emphasis on the crane has, however, had very beneficial results since it has focused attention on the need to plan the work in some detail before work starts on the site.

Some considerable improvements in overall site productivity have been obtained where tower cranes have been used. At first these improvements were credited entirely to the tower crane itself; later it was realised that two changes had taken place together, the introduction of the tower crane and site programming. In the course of efforts to maximise crane output various operations were studied in detail and were reorganised around the tower crane and the operations it performed. This introduced the use of work study techniques on a greater scale into the industry and a range of techniques was introduced to reduce the overall site time. For example, materials were carefully stacked in order of use and convenient to the reach of the crane, and much double handling was eliminated. Much greater attention was paid to the ordering of materials and to their delivery in the correct sequence. Experiments were made with units prefabricated on the site so as to use the full lifting capacity of the crane and to reduce the number of movements necessary.(71) At the same time, special equipment was devised to extend the range of things the crane could lift and to improve its operation. These developments took the form of special cages for bricks, skips for mortar, and remote control gear for operating the crane from the position which gave the best view of the operations.(72)

The increasing use of mechanical handling devices for materials on building sites has had consequences for both manufacturers and distributors of building materials. For example, the packaging of bricks, blocks and roofing in bundles of a size suitable for handling by clamp-barrows, hoists and cranes, not only reduces double handling and handling single units but also reduces losses and breakages. The availability of cheap polythene sheeting provides a ready means for protecting packaged units and prefabricated assemblies, and for bagging loose materials, keeping them dry and clean as well as facilitating handling, storage and recording. This not only assists the builder in handling and reducing wastage but assists distributors with storage, handling and recording. In order to make a full use of the possibilities delivery trucks are frequently fitted with mechanically handling devices.

OTHER TYPES OF MECHANICAL PLANT

There are also various other types of plant used in the construction industry, most of which fall into the category of powered tools, for example, compressor tools for breaking and drilling, and powered hand-tools for cutting, drilling, sawing and hammering. Generally these speed up the work of craftsman and labourers in particular operations without having a very great impact on the organistion of the work. In a separate category are the road-making plant. Road-making plant, spreaders, surfacing machines and rollers tend to be used at a high level of efficiency because usually they have a free run of work without being tied to the speed of the following operations. Within the family of building and civil engineering plant is the plant used for loading loose materials and for excavating minerals from pits and hillsides. There is also a range of plant developed for the smaller builder designed for versatility. For example, machines have been developed which can be used as a dozer, as motorised skip, or as a shovel. These machines are lighter than the full-scale specialist versions and lack their strength; however, they provide machines which can be used in a number of ways and hence with some chance of being used more frequently by the small contractor than would the specialised type of plant. It is, however, questionable whether the possession of such plant is more economic than hiring specialist plant or sub-contracting specialist work when the need arises.

COSTS OF MECHANICAL PLANT

Mechanical plant has, of course, a cost which must be set against the saving in labour and other costs which result from its use. From the point of view of a particular contract, the plant required is usually hired either from the parent firm or from a plant-hire organisation. The costs of plant to the contract are then the hire charges, the charges for transporting the plant to the site and away, the costs of any erection and movement of the plant on the site, fuel consumption, day-to-day maintenance, and the costs of the labour to operate it. With the exception of electric plant, plant is usually attended and its engine kept running during the whole working period of the operaion which it is assisting, whether or not it is productively employed. Thus the efficiency of site utilisation of plant depends on two factors, maximising the periods of use of the plant while it is on the site, and maximising the output of the plant while it is nominally working. However, as indicated above, overall efficiency is not necesarily obtained by maximising the efficiency of each separae operation. It does, however, indicate the need to try to arrange the operations for which the plant is needed, other things being equal, so that the plant is not kept idle for any length of time, and to examine the plant operations themselves to obtain the best balance between the various operations so that an output from the plant of something near its potential can be obtained. On average the output of plant per nominal working period is far below the potential, although some operators secure results much nearer the potential than others.[(68)] Improvements in operating efficiency can sometimes be obtained by choosing a single machine of average size to fulfil a number of functions and perhaps by meeting peak-load requirements in some other way. For maximising the efficiency of individual operations it is often necessary to have a range of machine sizes, but this tends to lead to each machine being used for only a short period and hence pushes up the machine costs. A better compromise is often to accept a lower rate of efficiency for the separate operations by choosing a compromise size of machine in order to reduce machine hire costs by raising the level of usage.[(73)]

Little precise information is available about the use of mechanical plant, but generally in Europe and America and other Western countries plant is used for mixing concrete, for excavation and soil movement, except where the job is very small. Powered hand-tools tend to be used to an increasing extent. Studies in Britain have indicated that for small jobs the mixer for concrete and mortar mixing and the hoist and powered barrows for handling materials tend to be only marginally economic for a large range of work. This is because they lack the flexibility of manual labour and tend to have a low output in relation to the period on the site. Nevertheless, they are more or less universally used because they reduce physical labour and make construction work more acceptable. For large jobs mechanical mixing is usually economic. Where a great deal of handling and transport is necessary it is usually economic to use the specialised plant available. Used in the right way cranes tend to be very economical, particularly as they are capable of both horizontal and vertical movements. Generally, mechanical plant is overwhelmingly cheaper than hand labour for digging, levelling and earth movement.

The direct savings in costs from the use of plant appear to be rather small unless the site labour is paid at high rates as compared with the factory labour manufacturing the plant. For example, figures for traditional house building in Britain suggest savings from the use of plant of no more than perhaps about 2 per cent of the cost of a house. Most of this saving would be realised from the use of a crane for materials handling and from mechanical earth movement. Such savings would be by no means automatic but would depend on the appropriate planning and organisation of the job. Mechanical plant lacks the flexibility of labour and its successful exploitation demands a high degree of planning to ensure that the plant completes its work in a short period of site time. Mechanical plant does not offer an easy solution to rapid increases in the levels of productivity but can be a valuable aid in this direction.

OTHER BUILDERS' PLANT

Builders have always used various types of static and other plant in addition to the more modern mechanical plant. Various improvements in this plant and in extensions to its range have been introduced which either save man-hours by being better fitted to their purpose than earlier forms, or enable building time to be extended by overcoming the difficulties of poor weather or darkness. Forms of scaffolding and formwork have been developed which are lighter and quicker to use, more foolproof and more adaptable. The development of cheap plastics has made it possible to provide cheap coverings for site operations which increase the range of weather conditions under which operations can be carried out. In Great Britain bad weather can result in the loss of 1 or 2 per cent of the time paid for by the contractor for house-building and as much again to operatives through being stood-off.[74] Losses can be much greater in places experiencing really unfavourable climatic conditions. For example, tents of plastic material either supported by scaffolding, or by pumped air, can be erected to cover the site even of quite large buildings. The introduction of steam curing of concrete increases the possibility of operating in frosty weather and of speeding up the curing of concrete and the turnover of formwork, so reducing the total site time. The development of profiles, jigs and mortar boxes reduces the need for skill and saves time in craft operations. Lighting enables work to continue during the hours of darkness. In Britain hours worked in winter can be 5 per cent less than those worked in summer as a consequence of fewer hours of daylight[74]. These and other devices all aid in reducing the time taken for operations and, by increasing flexibility and freeing the industry from some of its dependence on the weather, tend to reduce overall site time. But often the direct savings from the use of most of this plant are marginal; it is in the possibilities which they offer to the contractor to organise his work in the most efficient way where their real value lies.

CHAPTER ELEVEN

PROGRAMMING AND SITE ORGANISATION

THE INDUSTRIALISATION OF THE CONSTRUCTION PROCESS

The reduction of costs in manufacturing processes is usually achieved along three broad lines: by designing the product so as to reduce the work needed to produce it; by using machines which do the work more cheaply than labour; and by rationalising the productive process itself. In a similar way, the rationalisation of the production process is the third line of approach to the industrialisation of the construction industry.

The scope for reductions in the labour content of construction is thought to be considerable. Studies in Britain have indicated variations of labour expenditure for houses of as much as one to three.[(75)] Even when the extreme cases are omitted the range is still often very high.[(76)] The range does not appear to be reducing. It was estimated than nearly 10 per cent of the cost of site labour could have been saved had the below average firms increased their efficiency only to the level of the average.[(77)] Labour output depends partly on the way the work is organised and partly on the performance of the individual work-people. Organisation is mainly a problem of programming and programme execution, and the individual effort and output of the work-people a problem of supervision and incentives.

PRODUCTION PLANS AND PROGRAMMES

Unless a job is very small, and but few operatives are employed upon it, effective production planning will usually raise the level of productivity. Most building jobs tend to be very complex: even comparatively simple jobs may involve as many as a dozen trades and three or four times as many operations. These operations must be planned so that the labour gangs can move from one operation to another without delays. This needs carefully planning if all of the gangs are to be kept fully occupied.

The importance of maintaining continuity of construction is perhaps the first consideration. Continuity helps to reduce overheads and labour expenditure. Breaks in continuity tend to reduce the momentum of the operatives and to lead to periods when part of the labour force is left idle, or used on jobs at below their optimum output. There is no simple solution to the problems of keeping the operatives in continuous employment under optimum conditions. Moreover, the problem differs from one type of job to another because the number and type of operations and their length and the range of work-places differs from one job to another. One method of programming which has been devised is to develop the programme of work around the operation which takes the longest time.[(78)] For house-building this period has been found to be about 2-3 days. This period is taken as the cycle time and as far as is practical the gang sizes for the other operations are adjusted so that all the operations have a common cycle time. In this way a continuous programme can be planned for each gang so that there is a steady progression of gangs at each work place. Such a programme facilitates not only the movement of labour but also the ordering and placing of materials and machinery and the measurement of progress for interim payments and bonusing. In some cases reductions in labour expenditure of as much as a third have been achieved with the same labour force as a result of introducing a well planned programme.

It is, of course, unlikely that a programme can be strictly followed throughout the duration of a contract. Quite apart from the effects of bad weather, shortages of labour and materials and design changes, there is also a natural variability in the duration of similar jobs which results in queueing and in the breakdown of the programme. In fact, it would appear that with balanced gangs for each operation the contract will almost always exceed its programme time, balanced gangs being in many cases no more than a reasonable compromise in which it is equally important to finish the contract in the planned time and to save man-hours. According to this theory[(79)] gangs employed on the later operations should be proportionately larger than those on the earlier operations if the completion date is important and smaller if the economy in man-hours is important. Clearly, it is necessary to have rules to deal with departures from the programme as well as to have a programme. In the absence of satisfactory rules for dealing with departures from the programme, methods have been devised to deal with variability in operation times and to introduce slack into the programme so that departures resulting from delays can be made good.

The introduction of slack into the programme creates the risk that at times the labour force, or a part of it, will be underemployed. If labour is not to be laid off, and this is unusual in periods of heavy demand, alternative work must be found. Casual work tends to make a less than full use of the capacity of the workers. A common solution is to omit certain work from the programme which will be used as stand-by jobs. Such work must not be directly related to the programmed work and usually takes the form of auxiliary buildings, such as garages, boiler houses and small stores. The efficiency with which such work is carried out is likely to be less than that of the main work. Slackness in the programme is likely to lengthen the job and probably to increase the overheads.

Another method used for introducing flexibility into the programme is to limit the use of overtime to provide extra time for making-good delays in the programme. Of course, if much overtime had to be worked it might add considerably to the costs. The need for much overtime would indicate that the programme time was insufficient and that a longer time or larger gangs were necessary. In contrast, if the programme time was too long there might be considerable idle time.

A rather different approach to the problem of programming and a more radical one is to try to simplify the construction process and to reduce the number of separate operations. For example, it may be possible either to reduce the time for a particular operation and so bring it into a better balance with the cycle time or to eliminate an operation completely by prefabricating one of the components.(80) The prefabricated component might be produced on the site, at an off-site factory, or bought in from another firm. For instance, the operations of mixing and transporting concrete for over-sites might be eliminated by using ready-mixed concrete, or trussed roofs might be completely prefabricated. In the latter case a crane would be necessary. In this way it might be possible to either shorten or eliminate operations.

Generally, the fewer separate operations the easier it is to organise the work and lower the element of indirect time. A great deal of the time in building work is spent in the preparation for and in clearing up after an operation. For example, in some *in situ* concrete jobs as much as a quarter of the time taken for the operation is spent in preparation and cleaning up after the operation.(81) Much of this indirect time can be saved, particularly where operations take less than a day, if operations can be combined.

There are many ways in which operations can be eliminated or at least shortened by using prefabricated units. These may take the form of stock items such as joinery and metal goods, precast concrete goods, even prefabricated brickwork. Their use reduces the time taken and the complexity of the main structural process, partly by directly taking the work out of the main stream of activity and partly by reducing delays while *in situ* work sets and hardens. In this way breaks in the continuity of the work may be reduced and there will also tend to be reductions in the overall site time. Plant can be introduced in the same way to reduce time and bottlenecks. If this approach is applied to a particular design of building it may lead to the development of a system of building. In fact some of the systems have been developed in this empirical way.

Clearly, in determining the most efficient way to erect a particular building, costs must be examined as a whole. Any change in the process may set up a chain of reactions which will affect labour, material and plant costs, and overheads. It is immaterial whether a prefabricated component costs more than *in situ* construction as long as the overall effect of using it is to reduce costs. Again, a piece of plant is worth using even if it does not directly save sufficient in labour costs to offset its costs, if, over all, there is a saving. Many overheads are related to the duration of the contract, for example, the costs of site buildings, site supervision, other indirect site labour, and insurance; to a large extent the interest on the working capital is also related to duration of the contract on the site. While overheads tend to be reduced by shortening overall site time, this is accompanied by an increase in certain risks. The shorter the site time the greater the concentration of activity in the average day and hence the greater the disorganisation resulting from a day's bad weather, or from some failure to fulfil the production plan. With a heavily manned short programme both the risk of breakdown in production and the cost of meeting and correcting it will tend to be greater than where the same load of work is spread over a longer period. Hence the greater the concentration of the work load the greater the need to provide against breakdowns from any cause. Under such circumstances it may be wise to provide some protection of the work against bad weather.

The length of the programme which gives the most economic results will depend very much on individual circumstances. Clearly, there is an advantage in using each unit of production, whether it be men or machines, up to its capacity as far as this is possible without mutual interference, but the number of components of production is large and many have a large minimum size. It is, therefore, difficult to find the optimum programme period for a job. Sometimes, of course, the potential loss of profits to the building owner while awaiting completion will far outweight any possible savings from optimising the construction programme.

The system of programming by balanced gangs is only a simple first approach. More powerful methods of programme analysis are now available. These have been developed in America in other fields than building and are known variously as critical path analysis(82) and PERT.(83) The method consists broadly of listing all the operations which form the programme from start to completion. These are then plotted on an arrow diagram to indicate the order in which they are performed and the inter-relationships between the operations. This facilitates the study and improvement of the sequencing requirements. The next stage is to schedule the project by allocating times to the activities. First, the time of each operation is estimated, usually in days or half days, in terms of the duration expected with the gang anticipated to perform the operation. It is then possible to estimate the earliest and latest times at which each operation can start and finish without delaying the completion of the project and its overall duration. Not all operations follow or need follow consecutively and the spare periods are called float times. The critical path is the chain of operations which must follow consecutively and for which there are no float times. Any delay in performing a critical activity will lengthen the overall period of construction. A reduction in the time of a critical period may not reduce the overall time since the result may be to change the critical path.[84]

The effect of changing operation times can be traced through the diagram in terms of the effect on the critical path and on the overall contract time. Such a diagram can be used as a basis of costing, and in fact, it provides perhaps the best basis for estimating the costs of the contract since it gives the calendar times for each operation and not just the working times, so that allowance is made for idle time as well as productive time. On the basis of such a diagram it is possible to estimate the effect of combining operations, of substituting precast and other prefabricated work for *in situ* work, of introducing machinery and of other changes in method. The diagram will indicate the changes in the schedules of operations and, for example, the length of time each labour gang and piece of machinery is required on the site. It can also be used as a basis for preparing the schedule of material deliveries. The critical path method is a technique which can be used for optimising the solution to a planning problem; what is optimised depends on the user. The criterion might be minimum contract period; it might be minimum cost; or, again, it could be maximum certainty. It might, for example, be required to maximise the likelihood of completion on a certain day; in such a case efforts would be made to arrange for the ciritical operations to be those least likely to be delayed by the weather and other causes of delay. If there is a breakdown in programme the diagram provides the basis of determining the best remedial action. Clearly, the critical path diagram provides a valuable tool for the rationalisation of building processes; with its aid the study of new building methods can be carried a long way by paper analysis.

The tighter programming becomes, the more expensive departures from the programmme will be, and the greater the attention which needs to be given to causes of delay such as shortages of materials or labour, or delays from bad weather.

Operational research has also been used to develop decision rules of value in managing building operations.[85]

SITE SUPERVISION

Whereas programming and organisation is mainly concerned with the dovetailing of the various activities and with ensuring the proper supplies of labour, materials and plant, supervision is usually considered to be more a matter of obtaining the right quantity and quality of work from the operatives. In the past too much attention was often given to operatives' individual efforts and not enough to their organisation. It is now more and more accepted that output is at least as much affected by organisation as by the operatives' own personal efforts. Clearly, the output will be greater if the materials and plant required are to hand, if proper assistance is provided to the skilled craftsmen, and if the work is so organised that the gangs of labour can move and know they can move direct from one operation to the next. Again, a provision of adequate artificial lighting and protection from the weather will enable operatives to provide a better standard of work and to achieve a greater amount of work. In the light of present knowledge of site organisation and the operation of sites there is some doubt as to whether the effort expanded in developing systems of incentives is as valuable as was once thought. Nevertheless, of course, even the best-organised job will not be efficient if the operatives do not achieve the output and quality required. All planning of jobs must be based on assumptions about output per gang and no programme will run smoothly if the assumed output is not achieved.

The output of an operative can be increased in four ways:

(1) by reducing the operatives' non-productive time by for example better time-keeping, a reduction in idle time, and shorter pauses during the working day;
(2) by increasing the effort rate of the operatives to the optimum effort rate;
(3) by increasing the effort rate of the operatives beyond the optimum — this would eventually lead to the physical breakdown of the operative; and
(4) by a reduction in the standard of workmanship leading to work of an inadequate standard and to the wastage of labour and materials.

The first two methods produce desirable results and the second two methods produce undesirable results, although in some cases it may be more economic to waste materials than to waste labour.

Clearly supervision without some sort of sanctions is of limited value, since if operatives would do their best in the absence of sanctions then supervision in the sense used here is itself largely unnecessary. In times of a labour surplus the loss of employment may in itself be a sufficient sanction to back supervision; during periods of labour shortage other sanctions are necessary and these must generally be positive. While the moral sanction of supervision may of itself be weak, it is usually fortified when accompanied by example and the creation of a team spirit.

In small firms, one of the principals often either acts as a full-time supervisor or combines supervision with work at one of the crafts, usually bricklaying or carpentry and joinery. A survey of house-building in Britain indicated that labour productivity was higher where supervision was provided by a principal of the firm than by a paid supervisor. Labour expenditure was about 15 per cent less on contracts supervised by a principal of the firm than on otherwise similar contracts where supervision was provided by a salaried foreman.(77) The difference was, in fact, about the same as that achieved by paying incentive bonuses.

Much of the work of a contract, of course, is let to specialist firms who carry out the work of one trade, the building owner contracting with them direct or through his professional adviser, or the main contractor sub-letting some of the work to them. Some of this work is let on the basis of labour only, the materials and plant being provided by the main contractor. In other cases, particularly where the firm contracting to carry out the work is a manufacturer of the materials to be used, the trades-contracting firm itself sub-lets the actual work to another group of people on the basis of a fixed price for the work. Thus in one way or another the group of operatives who carry out the work are often paid a fixed sum for the work rather than an hourly rate, and thus in effect are working on piece rates, which provide the most direct form of incentive to achieve a high output. Generally, specialist firms are smaller than the general contracting firms and are more likely to have principals either supervising or working as a part of the operative team. For these and other reasons the specialist contractors probably achieve higher labour productivity than the larger general contractors. It was found that for house-building sub-contractors achieved the same output as the main contractors in otherwise similar circumstances with 15 to 20 per cent fewer man-hours.(77)

There was also some evidence from the survey mentioned above(77) that man-hours were lower for main contractors who sub-contracted a large proportion of their work, the reduction in the man-hours needed for their own work increasing as the proporiton of work sub-contracted increased. This suggests that the work remaining may benefit from the greater supervision which can then be provided and perhaps from the standards of output achieved by the sub-contractors.

In Britain in recent years there has been a considerable increase in labour only subcontracting, often referred to as 'the lump'. This has taken place within the trades normally carried out by the main contractor rather than through manufacturers supplying materials paying their fixers what are virtually piece rates. Thus it is trades such as brickwork, carpentry, plumbing and painting in which this labour only subcontracting has increased. The work subcontracted has thus tended to cover a large number of operations with an agreed cash figure, 'the lump'. Since the operations tend to be ones which are highly integrated within the building programme, the use of piece work rates for large groups of operations may result in difficulties in site management. The 'lump' gangs incentive is to complete the work within the 'lump' in the shortest calendar time so as to move on to another contract. This may lead to a poorer quality of work, a lack of adequate attention to safety and a wasteful use of materials.

Governments dislike the 'lump' system because of the difficulty of enforcing the payment of income tax and national insurance by itinerary workers. Further such workers are seldom properly insured and have no pension arrangements.

BUILDING OPERATIVES AND TRAINING

Building and civil engineering operatives are required for both new and maintenance work, and for work both on the site and in off-site workshops. The range of operations to be undertaken is very wide. No one type of operative could possess the whole range of skills. Traditionally operatives have been divided according to the material with which they work, such as bricklayer, carpenter, plumber and painter, or according to operational group such as scaffolder, roofer and plant operator. The skilled tradesmen are usually assisted by unskilled operatives, frequently designated as labourers who assist by preparing for the operation, serving the skilled men with materials and clearing-up after the operation. Labourers are found principally assisting bricklayers, tilers and plasterers; trades for which heavy materials need to be handled. Other tradesmen are often assisted by apprentices or trainees who carry out the less skilled work. Labourers also carry out most of the site works.

Skill is a matter of degree. Few building operatives are without any skill. Broadly the skilled operative has tool, material and operational skills[(88)], while the unskilled operative does not have tool skills and has the other skills at a lower level than the skilled operative. The worker designated as unskilled mixes and lays concrete, prepares mortar, lays drains and paths, operates simple mechanical plant, handles materials and carries out other operations which require some element of skill.

Skilled workers vary considerably in the degree of skill which they possess. Traditionally the skilled worker served a long apprenticeship which gave him the skills necessary to carry out all and any operation related to his material. He was expected to know the capacities of his material, how it could be fashioned, used and fixed, the sequence of sub-operations, tolerances and so on, and be able to work on the basis of a general instruction, supplemented by a drawing. Such workers are the truly skilled.[(89)] From their ranks would rise the supervisors and master builders. It was such men who masterminded the construction of the medieval cathedrals and who were responsible for the carving and pargetting. It is the same type of craftsmen who today restore the fine buildings of the past. Today the high price of labour encourages designers to achieve their decoration through the natural texture of materials and colour and contractors to prefabricate or purchase prefabricated units. For most craftsmen the range of skills which will be demanded of them is likely to be very limited and many do not think it worthwhile to acquire a range of skills which they are unlikely to use.

Semi-skilled workers generally lack the wide understanding of the capacity of materials and of the techniques involved in their use. They require detailed instructions to carry out operations and can generally only carry out the simpler and more common tasks. Increasingly they limit their activities to one type of operation such as fixing plasterboard or insulation material.

Today even skilled operatives specialise in a narrow range of common tasks, especially in new building and civil engineering work. Much of the demand for skilled operatives does not require the ability to carry out complicated work; it is sufficient to be able to build simple brickwork, fix timber frames, floorboards and roof trusses or lay plastic tiles. Many firms, or departments within them concentrate on small dwellings requiring only a limited range of skills. The increasing use of prefabricated and pre-cut assemblies has changed the emphasis on sites from fabrication and assembly to fixing complete units. Specialist contractors, especially those related to manufacturers and distributers of materials, fix only a limited range of products. At the same time, however, operations are closely related and much time can be lost if several separate sets of tradesmen are needed for each operation. If operatives no longer need to be masters of every aspect of their craft, they may have the capacity to acquire skills ranging over several different crafts. At least some of the breakdown between skills might be across related types of operations rather than concentrated on one type of material.

Traditionally craftsmen in the construction industry serve an apprenticeship. The extent to which apprenticeships are served and their nature was revealed by a survey in Great Britain in the mid-nineteen sixties (*Building Operatives Work* - H.M.S.O). It was found that most skilled construction workers served apprenticeships in Scotland, the proportion declining to about three-quarters in southern England. The serving of apprenticeships was far more widespread in some trades than others. Many of the apprentices were granted day release to attend courses at technical colleges but a considerable proportion did not continue to the end of the courses and gain craft certificates. The range of work in which experience was gained on the job was generally limited at best to basic work. Usually there was no training officer responsible for craft training, or to give instructions other than through the work carried out. The courses at the technical colleges were often found to be too academic to interest the apprentices. Britain generally provides less training in crafts than other western countries, particularly than Germany.

Even within the construction industry where technological change is not very rapid, the skills needed are changing. Operatives need to be able to adapt, not only to new skill requirements and to working with new materials but to applying basic skills to different types of work as the weight of building construction required shifts from one type of work to another, between different building types and forms building and civil engineering and new and maintenance work; see Chapters twenty and twenty-one.

Fluctuations in building demand tend to be a disincentive to contractors to provide on-the-job training. When demand is declining and pricing is very competitive reducing training is generally one way of cutting costs and is justified on the basis that fewer craftsmen will be needed in the future. When demand is rising the rush to complete work reduces the time craftsmen have to train apprentices. Many contractors do not provide any formal training, or at best in other than basic skills. Some contractors rely on being able to bid away craftsmen which other firms have trained, reducing the incentive of firms to provide training schemes. Again the higher the wages of trainees, the less the incentive to train. The levies often required from contractors not providing adequate training schemes are probably not sufficient to persuade firms to set up their own schemes.

Both trainees and firms would probably gain from pre-site training in freely transferable skills. These would provide a basis not only for follow-up training in college or on the job but for subsequent retraining as the demand-for secondary skills change. One-level training is being questioned. Not all trainees wish or are capable of acquiring the higher levels of skills needed for the very skilled work and for organisation and supervision of other workers. Most will never undertake other than the most straightforward work. However, as indicated above, some might find it worthwhile to acquire cross-craft skills so that they could tackle operations requiring several traditional crafts.

LABOUR INCENTIVES

Clearly, the principals of medium- and large-sized firms are not available to perform the duties of site supervision and some alternative sanction must be provided. Direct piece-work rates are used in some countries but this arrangement is not always acceptable to the trade unions. Often the unions insist that the full rigours of piecework rates shall be softened by being underpinned by some minimum level of wages below which the operatives cannot fall. This attitude is perhaps most easy to justify in an industry, such as the construction industry, where a large number of the factors upon which site output depends are outside the control of the operatives, for example, the weather, the supply of materials and the proper integration of the operations so that operatives in following trades are not held up. Often, as in Britain, the minimum level of earnings is taken as the normal hourly wage rate, although, of course, a lower level would provide a higher incentive, particularly at low levels of productivity.

In some countries, for example, in Britain, incentive bonus schemes have been the subject of negotiation between the two sides of the industry. Generally, however, incentive bonus schemes have not found great favour, although they are perhaps more used in Britain than, for instance, on the Continent, where it would appear that plus rates and modified piecework rates find more favour.[86] Even in Britain, only about 14 per cent of wage-earners in the construction industry were found to be paid under payment-by-results schemes (the 'lump' apart) as compared with 42 per cent in manufacturing industry.[90]

It is likely that a large proportion of operatives receive extra payments of one kind or another. The mere existence of some type of extra payment, particularly during periods of labour shortage, by raising the "take-home pay" forces other contractors to raise "take-home pay" by a similar amount in order to secure labour. Often these extra payments are not related to output and take the form of flat extra hourly or weekly payments. However, plus rates are sometimes paid as individual merit payments, although, as far as is known, this system is confined to the smaller firms where the principals are in a position to assess the merits of the individual operatives. Lump sums not directly related to output are sometimes paid as a reward for good work. Some countries, for example, Norway and Denmark, make a considerable use of modified piecework rates. Elaborate schedules of rates are negotiated between employers and workers, through their unions. But provision is made for guaranteed payments in the event of earnings falling below normal wage levels.[76] Often payments are made to compensate operatives not included in the formal schemes.

Incentive schemes have been found to achieve savings in man-hours of the order of about 15 per cent;[77] the savings secured by extra payments appear to be somewhat less, because, no doubt, a large proportion of such payments are unrelated to output in any way. The savings in labour costs are less striking since part of the saving is absorbed in payments made under the schemes and in the costs of running them. The net savings after payment of the bonuses were found to be less than one per cent of the cost of building in the case of housing.[77] The cost of operating the scheme had to be met from this saving and from any resulting savings in overheads. No direct savings appeared to be made by the use of extra payments. The use of incentive-payment schemes often involves a great deal of expense, particularly for measuring up work and for calculating the payments. Some of the measurement may, however, be of value in organising the work on the site, in determining interim payments, and in planning and organising future work.

The form of target scheme often varies considerably. The most common type provides for a payment proportional to the saving, although usually less than 100 per cent of the saving is paid. Incentive schemes can also be arranged so that payment is made only if a set target level is exceeded. The latter type is probably less acceptable but may provide a greater incentive. Perhaps the most important qualities of a target scheme are that it should be direct and easily understood. It is difficult to target individual operatives as they tend to work in small gangs, but clearly the gang sizes should be as small as possible so that the effect of each operative's labour on his bonus earnings is as direct as possible. In fact, it has been found that productivity tends to decline as gang size increases both where bonus is and is not paid.(90) Furthermore, the smaller the gang size the simpler the scheme is for the operatives to understand. Schemes usually provide a greater incentive when the operative can check that the bonus is correct. To this end the target stages should be visual and perhaps not more than a week's duration. While the greater the proportion of saving paid, the greater is the chance that the operative will be satisfied—if bonuses are earned too easily they may not provide sufficient incentive.

Many criticisms of payments by result schemes are made. It is often claimed that they do not ensure that the greater productivity is made in the right way and that it is, for instance, necessary to provide a greater amount of supervision to ensure that the standards of quality are maintained. Many complaints are made of the working of such schemes and of the number of disputes which ensue. Contractors claim that operatives become discontented if they do not receive regular bonuses and that these are difficult to provide strictly within the schemes because of the inevitable variation in the time taken even for similar operations, and because of the difficulties in maintaining a smooth productive flow. Other critics consider that the savings achieved are too small to be worth all the complications which ensue.

Furthermore, output is a function of management and it is often argued that it is for management to lay down the level of output in relation to the demands of the programme. This argument has greater strength in manufacturing industry where the system of production is highly integrated and the machines set the speed of flow. As construction becomes more industrialised it approaches the condition where the rate of output must be determined in the light of the whole production process. Where the programme of production is fully integrated too much output may be as unwelcome as too little. The problem of organisation is then to secure the rate of output required by the programme of work.

If no direct incentives are provided it is particularly important to make a positive attempt to create good staff relationships. One of the difficulties in creating and maintaining good staff relationships in the construction industry arises from the amount of casual labour employed. It is difficult to create good relations between labour and management when the rate of labour turnover is high. To some extent a high turnover of labour is unavoidable when the locations of work change constantly from one part of the country to another. Local and regional firms are in a better position to build up a largely permanent labour force. Some firms try to build up permanent and selected labour forces. Instead of target bonuses they pay merit bonuses and a bonus for length of service, accompanied perhaps by pensions and profit sharing.

Good welfare conditions and safety precautions, both fields in which considerable improvements are generally possible, are also thought to be valuable aids to good labour relations. None of these arrangements provide a direct incentive to high output but may have as much or more effect than incentive payments in securing the high stable level of productivity required for industrialised building. Such methods operate by creating contentment and staff loyalty and so encourage the operatives to give of their best, and negatively tend to make the holding of a job with their own firm sufficiently important to replace to some extent the sanction of a loss of employment, absent in a full employment economy. Systems of co-partnership between the owners and workers in a firm are another way of creating mutual interests and staff loyalties.

CONTROL OF MATERIALS

Building materials usually account for about half the cost of building new developments, although rather less for maintenance work. A substantial volume of materials are wasted on the site. In the case of housebuilding it is estimated that as much as 10 per cent are wasted.[91] Wastage arises through such causes as damage, burying, theft, misuse, using incorrect materials, negligence in cutting and preparing and over-specification. For example, facing bricks are used where commons would be satisfactory, too much cement is used in concrete, too much material is mixed, trenches over sized and useful materials are back-filled or burnt.

As pointed out earlier, packaging materials is an aid to reducing losses from breakages, weather damage, waste and theft. However, controlling wastage is the responsibility of site management, requiring careful planning of the siting, stacking and protection of materials, supervision of materials handling and operations and adequate site accounting procedures.

PART III

The Construction Industry

CHAPTER TWELVE

ORGANISATION OF THE CONSTRUCTION INDUSTRY

CONTRACTS AND CONTRACTING

As explained earlier, buildings and works are either designed to meet the client's special requirements (bespoke building) or they are built by contractors for sale (off-the-peg building). In some respects the differences between these two types of building are narrowing. Off-the-peg housing is often sold before erection is commenced, perhaps minor modifications being made to meet the client's requirements. Usually shops and offices which are built speculatively are only constructed as shells, the fitting being left until a client is found and his wishes are known. On the other hand, many contractors have developed standard buildings which they will erect on the client's site and to which small modifications can usually be made. Thus, in a sense, a third way of building has developed which is in some ways distinct from bespoke and off-the-peg building. To some extent this form of building resembles the normal arrangements for the manufacture and supply of expensive capital goods. The manufacturer is responsible for the design, materials and manufacture, and offers a standard range of products which he will supply to order. The growth of system building is, of course, leading to the expansion of this type of building. In this way the three groups necessary for the creation of a building, designers, producers and constructors, are combined under the control of the contractor.

The designers consist of architects, engineers and surveyors. They traditionally design to the direct requirements of the building owner and act as his professional advisers. The manufacturers of the materials, in which are included for this purpose the related extractive industries, often also produce materials for other industries in addition to building. For some components the manufacturers also design the installation as well as manufacturing the components; sometimes they also contract for their fixing. The contractors, of course, carry out erection and repair work of a wide variety.

Normally, in developed countries, a building owner will appoint a professional consultant to supervise new building work or important alterations. The professional consultant will be responsible for the design and specification of the work, for the contractual arrangements and for the supervision of the contract. For minor works the design is sometimes left in the hands of the contractor, as it is, in effect, where the contractor supplies buildings to his own design. In such cases, where the contractor supplies off-the-peg building, or provides both design and construction as a complete service, there is usually no independent professional consultant to advise the client. Such methods are largely used in the supply of private housing and for some industrial and other types of buildings. Similar methods are also used for the supply of specialist engineering plant, for example, gas retorts and coke ovens. However, it does not follow that because both the supply and the design of the building is in the hands of the contractor, that there is no place for the professional consultant. A consultant can still be employed to advise on the choice of design and contractor, and on the contract and its implications. Usually a contractor supplying design and construction services will either employ designers on his staff, or engage them for his design work. In this way the design may be more thoroughly considered and integrated with the erection problem than would be possible where the design process was geared to producing only a one-off building. The difference lies not in the process of design but in the absence of the independent advice which the professional consultant provides to the client. Public and large private building owners usually have professional design staff to handle the purchase, commissioning and management of buildings.

In the case of bespoke buildings the designer usually arranges the contract and negotiates with the contractor on behalf of the building owner. On large or important jobs he often employs a clerk of works as his site representative to ensure that the building is erected according to the design and specification. There is no reason why a professional consultant should not perform similar functions where the contract is for design and construction. Indeed, even where a consultant is employed to design the buildings usually a large part of the design is provided on a basis of design and construction. The engineering services and sometimes the frame being left to a specialist firm who contracts to design and construct. Where the building is already erected at the time of purchase the part which the professional consultant can play is inevitably more limited but he can still advise on the quality and price.

COMPETITIVE TENDERS

For bespoke building the usual procedure is for the design team to prepare the drawings and the specification, and in some countries, certainly those following British practice, a bill of quantities. These, together with the agreement between the building owner, designer and contractor, usually form the contract documents. Normally the contract is awarded on the basis of a competitive tender; sometimes the contractor is nominated and a price is negotiated.

The difficulty in choosing between a group of competing contractors lies in determining which is offering the most competitive price. The prices quoted by different contractors will differ for a number of reasons:

(1) because of differences in efficiency and therefore in costs;
(2) because some firms will be more anxious for the contract than others, perhaps because they are short of work, or for purposes of prestige or goodwill, and hence they will consider it worthwhile to accept only a small contribution to their overheads and profits from the job in question;
(3) because some firms will have achieved a better understanding of the technical nature of the job than others and are prepared to reflect this understanding in their tender price;
(4) because some firms aim at lower standards of work than others; and
(5) because some firms will have found ways of ensuring a high final settlement price, even on a low tender price.

Firms quoting low tender prices for one or for a group of the first three reasons are likely to provide better value for money than those quoting low prices for one of the last two reasons.

To some extent differences in prices arising from the last three factors all illustrate failures in the nature of the contract documents. Generally, the drawings and specifications are the responsibility of the architect and their design consultants, while any bill of quantities is the responsibility of the quantity surveyor. The latter document lists, usually by trade, all the materials and labours with the quantities required to carry out the work specified by the designer. Often, however, the drawings are not very detailed or detailed drawings are not available when the quantities are taken off or even when the contract is signed. In such cases it is not possible to be certain what work is actually implied until it has been started and the detailed intentions of the designer and client are known. When the design detail has not been fully worked out at the time of tender, the bills of quantities can only be provisional; the tender price can relate only to the job as defined in the bills of quantities and other documents. Prices are placed against each item of work; these unit prices are for use as a basis of adjusting the settlement price where the work done differs from the work listed in the tender documents.

There is much criticism of the above procedures. It is claimed that the method is so coarse that the tender price is no more than an average price which fails to reflect the ease or difficulty of carrying out the work. Designers claim that it is useless to design for easy erection if there is no reflection of this in the price. Until designers have confidence that prices will reflect erection economies they will not in general pay particular attention to design economies of this kind and, as a result, prices will be higher all round than they need be. It is claimed that the only way to approach costing with any degree of accuracy is to work out the method of erection, the plant that will be used and the programme of work. Contractors consider that this approach is not possible under the conditions of tendering; the information is in the wrong form, the time allowed is insufficient, and the method proposed is too costly in view of the number of tenders made in relation to the number obtained. In practice, the contractor rarely builds up the tender price on the basis of the bill items. Where British practice is followed and a rate is required for each bill item these are often worked out from the tender price after it has been estimated in some other way. Sometimes the tenderer does no more than apply a unit rate in terms of floor area or cube, in which some allowance is made for the quality required and for the overall difficulties of construction. In other cases a rather more detailed approximate estimate is made based on previous experience. Where no bill of quantities is provided the contractor determines the extent of the work himself or employs some outside firm to measure the work for him.

It is difficult to define quality in the contract documents and so almost impossible to enforce any particular standard. This difficulty is increased because much of the work is rapidly covered up and poor workmanship does not reveal itself until some time after the contract is settled. Frequent and detailed inspection is necessary to check such abuses as dirty cavities, poor quality subfloors and poor and badly prepared paintwork. Thus there is plenty of opportunity for the contractor to keep down costs in both labour and materials at the expense of quality should he wish to do so. In the long run poor quality work results in a bad reputation with the professional side but usually there is no formal machinery by which others can be informed of unsatisfactory firms. Nevertheless, there is no reason to assume that building contractors as a whole are dishonest or have lower standards than any other part of the business community.

As stated above, contracts are awarded mainly on the basis of total price; the purpose of the unit rates is to provide a basis of adjustment where the work carried out differs from the work listed in the contract documents. Frequently there are such differences either because full details were not available at the time the contract was signed, or as a result of changes in design during the course of construction. It is here that advantages of procedure can be taken to obtain a high settlement price in relation to the original tender. The method is to gamble on those items which are expected to be subject to variations. Low rates are put in for items expected to be reduced in quantity and high rates for those expected to be increased in quantity. For example, foundation work may be put into the bill on a nominal basis as until excavation starts the depth and design of the foundations are often uncertain. Often the amount of foundation work is much greater than given in the bill. A high unit rate for foundation work then both increases the final settlement price more than proportionately to the extra work and ensures that the interim payments in the early stages of the contract are disproportionately large. In comparison, low rates for the finishes will tend to result in less than proportionate savings, if, as often happens, their quality and quantity are reduced, and at the same time will tend to further increase the proportion of interim payments payable during the early stages of the contract.

On the other hand, the bill of quantities does have a value as a contract document. As far as possible it documents the work to be carried out and provides an agreed reference for disputes. The unit rates provide an agreed basis for price variations. Its misuse by contractors is often a result of incomplete and faulty specification and drawings in the first place, and variations in the work required during the course of construction. There would be very little opportunity for taking advantage of the contract conditions by contractors if the drawings and specifications were complete at the time the work was put out to tender and if there were no changes in the design subsequent to the signing of the contract. To some degree it is reasonable that variations should be charged at higher than proportional rates since in upsetting the contractor's programme of work they usually increase costs disproportionately.

However, even under the British system of bills of quantities it is rarely that all the work is listed in detail. Usually the engineering work and the services are specified in general terms and only a provisional sum is inserted. In some countries, for example, Scotland and South Africa, the engineering work is treated in the same way as the building work and the quantities are taken off. Often, however, the engineering work and the work for the services are subject to a design and construct type of contract. The contractor for this work is provided only with a general specification and himself designs the installation and quotes for its supply and fixing. In doing this he prepares his own working drawings and takes off his own quantities. Neither these nor the details of the pricing are usually submitted to the client or to his professional advisers.

Of course, even under British practice not all contracts are based on bills of quantities. Smaller jobs are often based only on drawings and specifications. In such cases there is more room for argument as to exactly what the agreed price includes. Where there is no outside consultant there are usually no formal drawings and specifications and hence little contractual basis upon which claims can be argued. The less the work to be performed is documented the more difficult it is to compare one tender against another.

Contracts may also be based on schedules of prices to which are sometimes added rough quantities. The latter are only to set the scale of the job which is eventually settled on the basis of measuring up the work actually performed. Sometimes prices are written into the schedule by the client's consultant, the contractor tendering on the basis of a percentage up or down. The advantage claimed for this type of contract is that it enables work to start before the plans are complete. Where this happens, the programming and efficient control of the job is made more difficult and prices are likely to be higher. Difficulties also arise in measuring the work, much of which is inevitably rapidly covered up. In the absence of detailed plans and specifications there may be some uncertainty as to what is required and the unit rates may be set on the high side to ensure their adequacy. Perhaps this type of contract is of most practical value in the case of maintenance and conversion work where it is often difficult to determine what work will be necessary until the work is started. This is particularly true in the case of long-term contracts for the maintenance of buildings for which there is little choice other than a schedule of rates type of contract.

During periods of price fluctuations the term "fixed price contract" has only a relative meaning. Contracts usually contain a rise-and-fall clause which allows the contractor to charge the client the increase in the actual labour and materials prices, or one or the other, over those in force at the time the contract was arranged; for practical purposes the crediting of falls in prices does not often arise. Not only does the rise-and-fall clause introduce uncertainty into the budgeting of the building owner but it reduces the incentive to the contractor to keen buying of materials and to resistance to increases in labour costs, and thus in the long run tends to raise the level of building prices. But during periods of steeply rising prices the contingency allowance which would be added to cover this risk in the absence of a rise-and-fall clause might raise costs to the building owner as much as the rise-and-fall clause itself.

In addition to the types of contract described as fixed-price contracts, there is also a class known as cost reimbursement contracts. The general basis of these is a cost of works payment for labour, materials and plant, together with a fee to cover overheads and contractor's profits.[90] The simplest form of this type of contract is "cost plus", where the fee is calculated as a percentage on the costs of the work. This form provides a direct incentive to inefficiency, since the higher the costs of the work the greater the fee. An alternative to the cost plus type of contract is the cost of works plus a fixed fee for management which is based on the estimated cost of works. While this provides no direct incentive to either efficiency or inefficiency, it does provide a yardstick to efficiency in the comparison of actual with estimated costs. Lack of success will tend to be reflected in the contractor's reputation. There are some firms who do all their work on this basis. A third form of cost reimbursement contract is the "value cost contract". Under this type of contract the fee is fixed in relation to the estimated costs of works, but the percentage of the fee is increased as the saving in actual, as compared with estimated, costs rises, and reduced as the excess over estimated cost increases. This form of contract provides a direct incentive to the contractor to keep costs down. The difficulty, of course, with all cost reimbursement type of contracts is the necessity either for complete confidence by the building owner in the integrity of the contractor, or for the careful scrutiny of all accounts. Detailed scrutiny may easily absorb any savings this type of contract may bring. The bid in the case of this type of contract is in terms of the percentage fee.

The value cost contract often comes very close to direct labour. The difference is that under this type of contract management is purchased as a service, instead of being provided by paid labour. As a result some of the risks of contracting are still hedged by the client. Once the client employs all the labour direct, including the managerial labour, he accepts all the risks of contracting. It is difficult to find any category of costs which can be avoided by the use of direct labour other than the risk profit on the capital employed, and the savings claimed for this form of organisation are often either illusory, or a result of the exploitation of management personnel. Often, while the direct costs of labour and materials are fully charged to the job, not all the overheads are allocated. Often no or only nominal charges are made for land and buildings, for the value of money provided as liquid capital, for top-level supervision, or for services rendered by other parts of the organisation to the direct labour division. Direct labour is most likely to be economic where there is a steady flow of work, especially where the work is highly specialised and where the location of the work is rather remote, so that a great deal of travelling would be necessary for outside contractors' labour. The principal difficulties in obtaining an adequate level of efficiency lie in maintaining a steady flow of work and in achieving comparable output to outside firms within the closed and non-competitive atmosphere of a single organisation. The economies of the direct labour building department are similar to those of the non-operative departments of a building concern, which will be discussed later. The difficulties of a regular programme of work can be avoided to some extent by putting the fluctuating balance out to contract. Some of the incentive from working in competitive conditions can be obtained if at least a percentage of the contracts are put out to competition with outside contracting firms. Of course, unless some contracts are awarded to outside firms it will be difficult to obtain serious bids.

Contractors often complain of the wastefulness of a large number of tenders being placed for each contract. Obviously, the greater the number of tenders the greater the cost to the industry for their preparation, but the greater the chance that the client will obtain a really competitive bid. A large number of bids tends to be of advantage to the client providing that the reason for the low price of the best bid is not because of low standards, or because of selective pricing of the unit costs as described earlier. But since, in the long run, the costs of excessive tendering for each job will be passed on in the prices, the number of tenders called for should not be unnecessarily large. Results obtained in Britain suggest that only a small number of tenders are necessary to secure a price reasonable to both sides; perhaps half a dozen may be sufficient. It would appear that the distribution of tender prices is very skew and that while the range of prices increases as the number of tenders increases, the minimum price changes only slowly as the number of tenders is increased.

TENDERING PROCEDURE

The standard method of obtaining tenders is to advertise an invitation to contractors to tender for the work. This method brings in any contractor who cares to tender, amongst them often contractors in whom the designer or the building owner feels little confidence. The general understanding under the system of open tendering is that the lowest tender will be accepted and the rejection of the lowest bids cause embarrassment. Furthermore, open tendering tends to lead to a wasteful multiplication of the firms tendering, the cost of which, in the long run, is paid by the clients. A common practice is to modify the procedure and to invite tenders only from a selection of firms acceptable to the client and his advisers. While this method reduces the cost of tendering to the industry and ensures acceptable tenders, it is subject to certain disadvantages. The number of real bids may be reduced by firms not wishing to refuse to bid and putting in cover prices when they do not require the contract. Where the contractors are limited in this way it will be more difficult for new firms to enter the field. Moreover, and especially when the approved list of contractors is very limited, the contractors will tend to become aware of each other's identity and the organisation of price rings will be simplified and made more worthwhile. Some of these limitations are overcome by a suggestion[90] that contractors should be invited to apply for permission to tender. A select list would then be drawn up from those applying.

In some ways the system of tendering itself stimulates price rings, particularly where the tendering is selective. In order to preserve goodwill the cover price must be sufficient to lose the contract without appearing unrealistic. The cost of estimating such a figure can be saved by asking a possible rival to suggest a figure. The advantages of some formal arrangement for obtaining cover prices and subsequently for obtaining compensation for the costs of tendering lead to the formation of a ring. The successful operation of such a scheme provides an almost irresistible temptation to share out the contracts at prices acceptable to members of the ring.

Clearly, the formation of such rings in building generally would be difficult, since the number of firms is so large. But the number of firms capable of handling the larger contracts is much smaller, especially in the smaller and remoter towns, where distance and the supply of labour provides conditions favourable to local firms. The larger and more specialised the work, the less competing firms there are likely to be and the greater the opportunity of creating a monopoly position. This is a danger which arises in the use of system building.

NEGOTIATED CONTRACTS

Of course, not all contracts are let competitively, some are negoitated. Any of the types of contracts which have been discussed could be negotiated but competitive tendering is more usual. In some cases competitive tendering is not really practicable because there is only one firm in the area capable of doing the work satisfactorily. In other cases a close relationship develops between a client and a contractor and all the building work is given to the same contractor. This type of arrangement often occurs with repair and maintenance work and with conversions and adaptations. For this type of arrangement to be successful the contractor must take a long-term view and provide a sound service at competitive prices, and the client must have an expert knowledge of building prices to give support to his confidence in the contractor. In such cases, and especially where the client's work is specialised, negotiated contracts may be advantageous to both sides.

Another motive for negotiating contracts is where it is desired to bring in the contractor at the design stage. This enables the problems of erection to be considered in the course of the design process, so that the work of erection is minimised in relation to the type of building required. This arrangement is made where it is considered that the savings from economies in erection are likely to exceed any advantages of competition. The contractor may, of course, have been selected in the first place through some type of competition, perhaps on the basis of a schedule of rates.

Negotiated contracts are most usual where it is important to obtain early completion of the building and works. Sometimes work commences with a price agreed for only part of the work, the prices for other parts being negotiated as the work proceeds.

Often a price is negotiated with a contractor who has satisfactorily completed a previous contract of a similar nature. Where the work is repetitive, for example, low cost housing, it is convenient to let further contracts to the contractor who has already successfully built one group. The price negotiated is related to the price for the earlier contract.

While the negotiated contract may be more expensive than the competitive contract, this is not necessarily so. Where work is regular and assured the contractor can often use his resources more efficiently, and may well be prepared to share the savings with the client to secure his loyalty. As long as some work is still let competitively there is a reference to which negotiated prices can be compared. Certainly in Britain open competition now enjoys far less support than in the past.[92]

SERIAL CONTRACTING

Serial contracting is a development of the system of negotiating further contracts with a firm which has successfully completed a contract for buildings of the same type. Under this form of contracting there is a legal understanding that there will be a series of contracts of a similar form when the first is completed. The exact conditions may vary from one contract to another; allowance may be made for an adjustment of prices in the event of a general change in the level of prices, or to allow for an improvement in productivity. There may be clauses to allow subsequent contracts to be withheld if performance is unsatisfactory. Generally, however, subject to safeguards, a contractor entering into a serial contract can plan on the basis of a continuous programme of work for some years ahead.

Serial contracting should result in lower costs to the contractor, some part of which should be passed on to the client. Lower costs should be obtained partly from higher productivity, partly from improvements in organisation and from special plant worth obtaining where there is an assured long period of regular demand. There is plenty of evidence to point to the improvements in productivity which occur with repetition; some of the savings are quite substantial.(58, 78, 86) Where the effective size of the job is large there is a large volume of work over which the costs of developing special techniques can be spread. The possibility of such savings should lead to keener pricing of serial than normal contracts.

While serial contracting obviously has a special significance for system building, its value is likely to be as great for more traditional work. Repetition is, of course, essential, but repetition does not necessarily entail monotony. Even on the same site repetition, if properly used, can have pleasing results. The serial contracts need not, however, relate to the same site but may apply to sites spread over a large area. There is some reason to believe that many of the design changes made for buildings designed by the same office to fulfil the same need are quite unnecessary. Serial contracting is equally applicable to civil engineering, for example, to roads and bridge works. Contract prices might be very much lower in these fields if contractors could price their work with the assurance that their special plant could be certain of regular and full use for many years ahead. The opportunity for serial contracting is not difficult to find in the field of public works and buildings, for which needs can be programmed well into the future.

BILLS OF QUANTITIES

As pointed out earlier, the bill of quantities is by no means in universal use. It is essentially a British development which is used for most substantial contracts in Great Britain. Its use has spread to most parts of the Commonwealth, but not to America or Canada, or to most of the other European countries, although it is used in Germany. Generally, however, the bill of quantities is not used as a contract document outside the Commonwealth or ex-Commonwealth countries.

Generally, outside the Commonwealth there is no separate profession of quantity surveyor, although there are some signs of the development of such profession in some Continental countries, for example, France, Holland and Denmark.[(93)] Whereas in the Commonwealth the quantity surveyor prepares the bill of quantities, negotiates the pricing of variations, and generally acts as the building accountant, in countries not following this practice, the architect, as the professional adviser, fulfils this function, but without the aid of a bill. For example, in America the tender documents consist of plans and specifications only.[(95)] Each contractor competing for the work takes off his own quantities in order to price out the work but reveals only the total price to the client's professional advisers. The documents provide a complete specification of what is required but the contractor is left to decide how to achieve it. The drawings are usually much more detailed than under the British system but the architect does not subsequently issue working drawings; these are prepared by the contractor. Perhaps as a corollary to this the architect generally leaves the contractor to make his own arrangements for materials and components, and for sub-letting works. In Canada, and to some extent in the U.S.A., the sub-contractor's tenders are deposited with a central organisation so that the architect is aware of the sub-contracts and so that the sub-contracts cannot be peddled around after the main contract is signed. While the architect has the right to examine the contractor's proposals his vetting is limited to ensuring that the proposals satisfy the specifications and plans on which the tender is based. The architect cannot usually disapprove the contractor's proposals because he dislikes the solution. This sytem is thus similar to that operated elsewhere for the engineering work.

This system involves the duplication of effort in taking off quantities as compared with the British system and will tend to push up the contractor's overheads, especially if there are a large number of bids for each contract: eventually, the extra overheads resulting will be passed on to the clients in the form of higher prices. In the absence of an official and accepted bill of quantities there is no standard from which variations can be priced, and it is all the more important that the work shall be completed as specified for tendering purposes, although variations frequently occur. On the other hand, since no simple measure of work is provided, and since full details of the design are available at the time of tender, there is nothing to impede the contractor from pricing in accordance with the way he proposes to build, and hence any design economies should be reflected in the price. The absence of a detailed specification for each component leaves the contractor free to find the best offer on the market rather than being forced to accept the product of a specified supplier. It is claimed for this system that it relieves the architect of the need to know the market for materials and components, and puts the task on the contractor who is in a better position to know the market and to bargain for competitive prices. On the other hand, the contractor is likely to consider only the purchase price and to give less weight than the architect to long term performance, and to maintenance and other running costs.

While the incentive to the contractor for all-round efficiency, particularly in buying and determining the best method of construction, is theoretically greater than under the British system, it is difficult to be sure that the client obtains better value for money, particualrly in terms of costs-in-use, that is, in the combined initial and running costs. It is probably more difficult to ensure quality and to obtain the required specification in the end reuslt than it would be if there was a bill of quantities.

The bill of quantities, as normally used under the British system, has come under considerable cirticism both from the design and from the contracting sides of the industry. The bill is said not to reflect the way the work is carried out and that hence it does not reflect the actual costs of construction. The work of construction is, of course, carried out by operations and site records are compiled in this form, which thus provides no basis of costing the measured work units given in the bill of quantities. For example, brickwork is carried out a lift at a time, including the formation of openings and the building-in of window frames. It thus provides no measure of the cost of forming reveals, cills or other similar items frequently listed in bills. Moreover, in the conventional bill, items of work are measured by quantity and grouped by trade irrespective of where the work occurs, or of the difficulty of construction. For these reasons it is contended that priced bill items do not reflect the costs of the particular job but only the general average of jobs. This situation tends to increase the risks of tendering, and hence raises average prices and discourages designers to look for design economies.

One solution suggested to this problem is the "operational bill".[95] Under this type of bill the work is not taken off in units of work, labours, but as complete building operations in the form in which the work is expected to be carried out. A building operation implies a continuous piece of work carried out by one particular gang and terminating where the work of a following gang starts. While the materials can be measured as in the conventional bill, it is though better to measure them also in terms of operations, so that they are in a suitable form to be arranged into schedules for ordering. The labour content is measured for the complete operation and no attempt is made to measure the times for each item. The time for the operation can thus be measured in terms of the gang of men and the number of hours or days they need to complete the operation. This can be estimated from the recorded site times for previous work, and at the end of the job the estimated times can be compared with the recorded times for the operation.

The possibility of an elemental bill has also been discussed. Under this type of bill, items would be grouped by building elements instead of by trade. This system would be convenient as a basis for analysing elemental costs and for preparing elemental cost plans, but it appears to have few advantages for the contractor.(96)

SUB-CONTRACTING AND TRADES CONTRACTING

It is unusual for one contractor to carry out all the work with his own labour, except when the job is very limited and simple. In some countries, for example England, America, the Commonwealth countries, and generally in Holland, all the work is let by the client to a single contractor. The contractor then sub-lets work to other contractors. Usually the main contractor carries out the main trades, masonry, concreting, excavating and carpentry with his own labour and sub-lets some of the finishing and the servicing trades. Sometimes the main contractor will sub-let virtually all the operations, he himself being responsible for organising the work. Often, the building owner, or more usually his consultant, will specify the firms to which certain operations are to be sub-contracted. This is a British practice and is not followed in all countries, for example, in America. Under the British system a provisional sum will be stated in the bill, or specification to cover the item. The main contractor's financial interest is limited to a percentage of this sum, but it is he who contracts with the specialist firm and who must accept responsibility for the work of the sub-contractor and for fitting his work into the organisation of the job. Such specialist firms usually provide the design and the materials as well as carrying out the fixing. Their use for heating, ventilating, lighting and elevator installations, and for structural steelwork is a convenient way of getting design work carried out beyond the competence of the architect, although there is some doubt whether the designs provided by those supplying and fixing materials are as economic as designs prepared by professional consultants. Contractors complain that the scale on which sub-contractors are nominated is sometimes so great that the organisation of the contract is jeopardised, since the contractor finds it more difficult to co-ordinate nominated sub-contractors than those sub-contractors who, because chosen by him, depend for further work on his goodwill and that of other main contractors.

In other countries, for example, Scotland, France, Denmark, Western Germany and Austria, the separate trades contracting system, or *corps de métiers* is the basic method employed.(86) Under this system the architect, acting for the building owner, puts out each trade to tender separately. The contracts are normally for labour and materials. The contractors are usually organised on the basis of a single craft, the exception being the general contractors who normally carry out excavation, masonry and brickwork, and sometimes also carpentry and joinery work. The number of separate contractors for each job is often quite large. For large jobs, even for housing, the number of separate contractors may exceed twenty, or exceptionally even thirty.

Perhaps the main difficulty with this type of contracting lies in the organisation of the work on the site. This is the responsibility of the architect but usually he has neither the experience nor the time to organise and supervise the site work. As explained earlier, the programming and orgnaisation of the site work is one of the keys to high productivity. It is not suprising that the periods of construction are generally said to be longer with the trades contracting system than with main contracting. A variant of this system is to appoint a leading contractor, usually the contractor carrying out the main constructional work, to co-ordinate the work of the other contractors. This tends to raise the costs of the job without solving all the problems. The trades contractors tend to be more independent of the leading contractor than they would be of a main contractor, since they do not look to him for regular employment.

The amount of trades contracting is probably on the decline. For example, in Scotland probably less than a third of the work is now let in this way;[(97)] this form of contracting tends to be retreating before competition from English main contracting firms and from the increase in the use of proprietary systems. Similarly, the use of proprietary systems is reducing the importance of trades contracting on the Continent. Another factor which is encouraging the growth of main contracting in some countries is a change in the system of taxation to a form which does not penalise the sub-letting of work.

Where, as in Britain, it is the practice to nominate the specialist sub-contractors, a large part of the contract is often settled without any competition. Sometimes as much as 50 per cent of the value of the contract is awarded to named sub-contractors without any competitive bids. This seriously reduces the competitive nature of contracting. In America, where there is no such limitation on the freedom of the contractor to sub-let work, the contractor has a direct incentive to secure the best prices for the work he sub-lets, since this reduces his costs and hence increases his ability to be competitive. Since he is the direct employer of the sub-contractors, he has direct control over them. Under this system the requirements for the engineering work must be specified much more closely than under the British system.

CONTRACT PERIODS

The client requires a low price, good quality and an early completion date. The speed with which the work is completed is often of considerable importance both to the client and to the community. The longer the period of construction the greater tends to be the cost of the capital tied up in the construction. Delays in obtaining the building often result in losses of profit to the client. When the urgency for a building is great little attention tends to be paid to economy of construction; often contracts are made on the basis of schedules of rates and construction commences before the design is settled. Such urgency is often a result of a lack of forward planning by the client. From the point of view of the community it is often the duration of the work on the site rather than the length of the contract period which is important. The duration of the site work is particularly of importance for work on roads and in other public places, where the construction work impedes the flow of traffic and adds to the costs of transport.

While better programming and improved site efficiency can result in a reduction in the duration of the contract and in a reduction in overhead as well as in prime costs, there is a limit to which the duration is worth reducing. The shorter the period of duration, the greater the concentration of work, and the greater the risks that costs will be seriously increased by delays resulting from bad weather or other causes. The duration of the work on the site is often lengthened because the site work is commenced before the contractor has had an opportunity to complete the site programme and to secure adequate supplies.

The size of contracts for new construction tends to be quite large compared with the size of most contractors' organisations. Moreover, as described earlier, under the system of competition there is very little certainty of obtaining any particular contract. It is, therefore, difficult to ensure an even flow of work for any but the largest contractors. Clearly, contractors do not wish to have to run down their organisations, particularly in times of full employment, but they cannot, of course, carry for long labour and plant for which they have no work. Contractors, therefore try to ensure their suppy of work by taking a fairly gloomy view of their chances of success in tendering. As a result, and particularly in times of full employment, they tend to become overloaded with work in terms of their capacity. Clients expect to see work started on the site soon after the contract is signed and, as a result, contractors tend to spread their labour too thinly over the sites in order to be able to show some progress on all their contracts. This reduces efficiency and probably results in the contracts taking even longer to complete. One remedy would be for clients to plan their work ahead so that contractors could look ahead to a long forward programme, when they would be able to plan their work as a whole as well as at contract level. Another remedy might be for the contract to contain a completion date but no starting date, and for a system of large bonuses to be payable for completion at stated times ahead of the final contract date, the times and bonuses being fixed in relation to the client's needs. This would provide an incentive to early completion while giving contractors an opportunity to plan their load of work. The client would gain particularly where use could be made of the units completed, such as dwellings, offices and sections of road.

THE CONSORTIUM

Building consortia have been created both by contracting organisations and by client organisations. Engineering firms have formed consortia to contract for building nuclear power stations and other engineering projects, particularly for work overseas. Clients have grouped into consortia particularly in connection with the ordering of system-built constructions. For example, in Britain groups of authorities have pooled their orders of housing and schools, and in some cases they themselves have developed special systems of construction, for example, the CLASP system.(67) In Britain there is considerable government encouragement to the formation of consortia by public authorities. Consortia can take several forms and fulfil a number of functions.

The point has been made earlier that where a considerable amount of special plant is needed for a form of construction, costs fall at first quite rapidly as the number of units over which their use can be spread increases. This is equally true of any technique, whether it employs plant or not, for which the capital costs of development are fairly heavy. It is equally true of traditional as of non-traditional types of construction. Techniques of construction and organisation which are expensive to develop may not be worthwhile if there is no certainty of their use for a considerable number of units of construction. Thus a long-term guaranteed market may enable a contractor to introduce specially developed techniques or special purpose plant which will greatly reduce the costs of construction. Moreover, a large-scale, long-term purchaser who is able to provide a guaranteed market of the right scale is likely to be able to obtain a share of the reduction in the costs secured in this way. If the contractor has to carry the risks of sufficient orders materialising, then, clearly, he will spread as much of the capital costs of development as possible over the first orders and himself secure a large share of the savings, should the flow of orders continue.

Generally, where the clients are only occasional purchasers of buildings, there will be no guaranteed forward orders. The risks of development will be carried by the developer and he will expect to retain any savings from the developments to compensate for the risks accepted. In such circumstances the development of new techniques may be retarded. On the Continent, housing and some other types of buildings tend to be purchased directly by large government or government-sponsored organisations who can offer orders of the size and continuity necessary to make the development of new methods worthwhile. This may be one of the contributory reasons for the earlier large-scale development of proprietary systems in some Continental countries than in Britain and other countries similarly organised. In these countries most of the public purchasing authorities are small, although collectively the total volume of building ordered each year is very large. Guaranteed orders can only be provided on the right scale if the purchasing authorities come together and agree on a standard design to meet their future requirements. Such joint purchasing organisations may be difficult to organise, at least, in the first place, when each authority has been accustomed to having buildings specially designed to meet its own requirements. A considerable measure of freedom must be sacrificed if a consortium is to be organised.

Consortia need not confine themselves to negotiating for proprietary systems. Equally well they can develop their own systems; they can employ staff to design a special system and the component parts for it, and can then enter into special arrangements with manufacturers to produce the components and with contractors to erect them, or they can set up direct labour units to handle the erection. The CLASP system was designed by the client organisation and it is they who designed the components and arranged for the manufacture of the units and with contractors for their erection.(67) Such a process need not be confined to a system based on large units but could be operated equally well with a traditional form of construction.

Clearly, consortia could be developed to provide a countervailing force to the manufacturers of components and materials, and to contractors. The consortium must, however, be large enough to be able to place orders of sufficient scale and continuity to make it worthwhile to set up units to supply them. This does not mean that, for example, production units would do no other work but that some sector of a firm would concentrate wholly or for long periods on orders for the consortium. The consortium could then use the power provided by its size to secure an equitable share in any reductions in the production costs resulting from the scale of its orders and from savings in sales and distribution costs. This, in effect, is the procedure of large-scale manufacturers and large-scale retail distributors in placing orders for supplies. Sometimes the scale of purchase is so great that the entire output is purchased by one organisation. Where this happens the manufacturer obtains considerable savings in the costs of sales but soon becomes completely dependent on the organisation to which the products are sold. The problems of distribution are normally greater with building than in other fields since the products tend to be bulky, heavy and comparatively cheap and cannot stand heavy transport costs. This tends to limit the area of the market, especially for complete buildings and makes it more difficult to build up a consortium of sufficient size.

The possibility of forming consortia is not confined to public authorities but is possible for any group who can provide regular orders of sufficient size, for example, co-operatives could be formed for purchasing farm buildings or industrial buildings. Perhaps, however, the greatest opportunities are in the public field, particularly for public housing, schools, health buildings and road building. However, a consortium will usually be at a disadvantage as compared with a single authority since the problem of compromise between the different authorities always causes difficulties. So far, at least in Britain, neither the large cities nor the new town corporations, the organisations with the largest programmes of construction, have shown any conspicuous success in securing the lower prices which the scale of orders might be expected to produce.

Consortia of clients could lead to consortia of contractors who could use their power to make good bargains for the bulk order of materials. This might, in turn, lead to consortia of manufacturers who could allocate orders amongst their members and further increase the scale of specialisation and the length of runs. Of course, if this happened the countervailing power of the clients and the contractors would tend to be reduced, leading perhaps to a more equal distribution of the gains from specialisation and scale of production. There is also a danger with consortia on the supply side that some element of monopoly may be created.

THE PACKAGE DEAL

Under the package deal type of contract a purpose-designed building is provided, the contractor providing both design and construction. Often the contractor will study the client's business, determine the nature of the building needed, design it, find a suitable site and erect it, all the services being provided for one lump sum. To the client the service offers a guaranteed price without any extra items, no organisational problems and a guaranteed completion date. This type of contract does not, however, reduce the number of functions to be performed; it only removes their control from the client to the contractor. Any savings achieved by the method must result from higher efficiency from the individual services and from better liaison between them. Often, however, the contracting organisation only acts as the developer, or as a constructor, and other services are provided by separate firms of consultants and contractors in the traditional way. The client, however, has no independent consultant to advise him on the adequacy of the designs offered or of the reasonableness of the price; the price, of course, is negotiated. Usually the contract does not closely specify what is to be provided for the price, so that even if a consultant is hired, at extra cost, it is difficult to evaluate the building in relation to the cost, or to determine whether the implied terms of the contract have been fully carried out. The success of the package deal type of contract, therefore, depends on goodwill. While the package deal does perhaps establish a situation in which co-operation between the design teams and the contractors is easier to achieve, it does not appear to go very far towards uniting the various parts of the construction process.

A development of the package deal is the 'turnkey' type of contract. Such contracts usually embrace not only the building and site works but the furnishing and equiping of the building and can include the hiring of labour and management, and arrangements for stocking and the regular supply of materials. The most far-reaching types of 'turnkey' contract have been signed with developing countries, such as Arab oil countries. For example, 'turnkey' contracts for hospitals have covered all type of staff and medical supplies. At the end of the contract the hospital has been ready to receive its patients. Contractors in many Western countries are now competing for such contracts but the volume of business is not as buoyant as it was.

DEVELOP AND CONSTRUCT CONTRACTS

Another variant of the package deal type of contract is the develop and construct type of contract. Under this the clients' professional consultants produce a detailed brief of the buildings required, often with site layout plans and layout plans for each type of building. Contractors then tender on the basis of their own type plans being free to offer their own design solutions subject to satisfying the specification and other details in the brief. This type of contracting is approaching the American method. It is particularly appropriate where, as in house-building, contractors have developed their own house-types and methods of construction.

INTEGRATED CONSTRUCTION ORGANISATIONS

In the past the designer and constructor were usually one, and much thought has been given to the problem of reuniting the whole structural process from design to completion within one organisation. Up to a point this is achieved by the contractor providing off-the-peg buildings, and to some extent by the suppliers of proprietary buildings. Clearly, if the designer is absorbed into the contractor organisation certain risks must be accepted. Aesthetic and user needs may be subordinated to the needs of erection. There will be no professional independent of the contractor to advise on the adequacy of the design from a constructional or user point of view, nor on the extent to which it provides value for money. The ultimate position might not be very different if the designer absorbed the building organisation.

As an alternative to uniting architect and builder, architects could themselves specialise in certain types of building and so acquire a deeper knowledge of design requirements in relation to production. If this approach is to achieve a considerable advance in design and production efficiency, it would probably be necessary for the designer to produce not only detailed drawings but the site erection programme with schedules for ordering materials and organising labour. The designer would increase his understanding of the relation between design and construction if he had a site management function for at least some of his design jobs. The contractor's normal role would then be limited to site and job management.

THE INFLUENCE OF GOVERNMENT ON CONSTRUCTION

Consideration has already been given to the ways in which the government influences the level of demand on the construction industry. In this section it is ways in which governments can influence the organisation of the construction industry and its products which needs to be considered. Such influences are exercised through bylaws, planning control and the incidence of taxation.

Governments influence construction through direct physical controls by the issue of licences and permits. Sometimes such controls are operated by the licensing of building construction, a licence only being issued subject to satisfying a government agent as to need on the basis of certain criteria, and usually carrying limitations on scale and quality. Control is also exercised through the issue of permits for the purchase of materials or for obtaining labour. Sometimes the methods are used in conjunction. Such methods are used either to reduce the overall amount of construction, or to limit construction for certain purposes.

The amount of construction work may be increased by government action either because they require more buildings and works to meet policy needs, or because they wish to stimulate the economy. A large number of aspects of government action generate a need for constructional work. Attempts to improve social conditions generate a need for housing, schools, hospitals, and other types of buildings. The improvement of transport and communication facilities generates a need for roads, bridges, harbours, and other similar facilities. Increases in population and changes in the forms of economic acitivity all generate a need for constructional work. Up to a point, requirements for buildings and works can be met from existing stocks; existing buildings can be adapted to meet current requirements and standards can be reduced. By direct and indirect intervention governments can stimulate the demand for buildings and works when conditions are opportune. The view is often taken that the construction industry is very flexible and that the volume of work can readily be changed to suit the needs of the economy as a whole. It has been suggested that in times of unemployment public works can be put in hand rapidly and thus provide an easy means of directly reducing unemployment and of pumping money into the economy, and so setting up a chain reaction which will spread to all industries; resulting in an increase in total economic activity greater than the initial stimulus to construction activity. This is what is called the multiplier effect. While there is little doubt that an increase in the amount of construction does directly affect the demands on a large number of industries, and thus provides a wide base from which to stimulate the economy, there is some doubt whether it is easier to expand output rapidly through the construction industry than in other ways. The planning of building construction and public works takes a long time. It involves long-term national and regional planning to determine the best location of the development at a broad level and town planning to ensure that the development is located to produce the most suitable land use pattern. The buildings and works must be designed and land acquired. Thus months and years of work are often necessary before work on site can start. Moreover, while the construction industry may employ a great deal of casual labour, it is not unskilled labour, and much of the less skilled work makes heavy demands on the strength and hardiness of the labour. It is not the type of work likely to appeal to, or to which the majority of unemployed would be suitable. Furthermore, as the construction industry becomes more industrialised, so the need increases for labour of special skills and for special management skills and capital equipment.

It has been advocated that programmes of public works should be planned, and designed and held ready to put in force when economic activity reaches the level at which the need for such drastic stimulus arises. While it is possible to plan future programmes of development, good value for money will only be obtained if they are planned in relation to need. If the facilities they are designed to provide are then created earlier, or postponed, they will be likely to provide less value for money than if they had been provided at the optimum time. Thus the buildings and works provided as a result of a public works programmes for stimulating the economy may be far less valuable than buildings and works provided to meet needs in accordance with a planned programme of urban development. It seems probable that the slow progress in providing roads and other urban facilities in the United Kingdom in the last two decades may have been associated with a policy of holding such work in hand until the need to stimualte the economy arose. The cost in terms of national efficiency of delaying building and public works during periods of prosperity needs to be set against the gain which might be obtained from stimulating the economy with a programme of public works should the need arise.

The effect of using the construction industry as a stimulus to the economy will effect the industry itself as well as the overall economy. Construction is a multifarious activity comprised of constructors working in different types of materials and on different types of constructions, and of a wide range of special trades carrying out engineering and finishing work. A cut in road construction does not release labour, management and plant suitable for erecting houses or hospitals. Electricians or floor finishers cannot be switched to building irrigation works. The imposition of controls on some types of building activity may do very little to shift resources in other directions. In fact, if controls are not applied with care, the result may only be to create sectional unemployment in some parts of the industry without easing pressures elsewhere. As a result, the national resources may be used less efficiently than before. This is not to say that government action may not be effective in moving resources in the required direction more rapidly and less wastefully than the action of uncontrolled economic forces, but that such government action is only likely to be successful if it is based on a very thorough understanding of the industry and the way it operates in the economy. Where the proportion of orders initiated by the government is large, many firms will have a higher proportion of government contracts and be at risk from changing levels in the scale of public contracts.

In some countries, notably those in Eastern Europe, the State tends to play a dominant part in the construction industry through its dual functions as *entrepreneur* and client. Long-term development programmes are prepared so that the construction industry and the associated materials industries can plan ahead for long periods. More detailed plans are prepared for the period immediately ahead. For example, in the U.S.S.R. each town has its own building organisation which is divided into separate trusts for each district and for specialised work, such as road works.(98) These trusts place large orders for work to a standard design. Such a system should stimulate innovation since the development costs can be spread over a large volume of work with a reasonable chance of achieving an overall reduction in costs per unit.

Governments tend to have a dominant role in construction in developing countries since the government and agencies are usually the main client for constructional work. In developing countries the constructional industry is of considerable importance since the development of the economy depends, to a large extent, on the provision of buildings and works. The danger is, therefore, particularly great of attempting more than the construction industry can handle and hence of reducing efficiency and raising costs without increasing the output of the industry. There is also the danger of stimulating the use of labour-saving methods when abundant labour is available either from unemployed or underemployed people. While in the circumstances of developing countries labour may be comparatively inefficient, it is usually cheap and abundant, so that its real costs are low in relation to the labour costs incorporated in plant and prefabricated components which might be substituted for it, but which would have to be imported from countries where labour was more expensive. Thus, in relation to the costs of labour, plant and prefabricated components are likely to be more expensive in developing countries than in developed countries. As a result, industrialised methods tend to be less economic in developing than in developed countries and will tend not to be worth using on the same scale. Situations will, however, arise in which, for example, plant is economic in developing countries, particularly where the works have to be performed in remote places and in climatic conditions under which human labour has a relatively low output. On the other hand, circumstances may arise under which it is better, from an overall national point of view, to use labour intensive methods at slight extra cost as a short-term policy, either in order to save imports, or because the economy will be more stimulated by spreading the spending resulting from the construction widely within the country. Often labour, especially rural labour is not fully employed and the cash economy is limited. The inhabitants do not have the money to pay to hire labour or the services of contractors but could meet their own building needs by carrying out simple construction operations in their spare time. Their need is often for outside expertise to design, provide simple instruction in building, organise and supervise construction work. This would be analogous to self-build in Europe.

As mentioned earlier, where governments cannot exercise the degree of control they require by fiscal and financial means they sometimes impose physical controls. These variously take the form of licensing building work according to priorities and licensing labour and materials. In fact, it is difficult to apply the same detail of control to the private sector as to the public sector without physical controls. Such controls are, however, difficult to apply without distorting the economy of the industry and reducing its efficiency. It is difficult to draft regulations to take account of the complexities of an industrial economy. Usually the regulations must be very broad and their interpretation left to be applied by government officials. Unless the very activities which the licensing is designed to assist are to be stultified, it is difficult to prevent less essential needs being licensed as a result of special pleading, while in default of similar astute presentation more essential needs may be turned down. As a result of these activities a great deal of ingenuity, better applied to construction problems, may be wasted in defeating the spirit of the regulations, while uncertainty and delay will tend to reduce the level of productivity and raise the level of costs. Another objection to controls of any kind is that they create windfall gains to those sectors of the economy in possession of the rights to the privileges generated. It is sometimes suggested, that where licensing is necessary, the licences should be auctioned to the government's advantage so allowing priorities to be determined in the normal way by market forces and eliminating windfall gains. However, if the priorities could be determined by market forces, it is probable that financial sanctions would have been sufficient in the first place and that physical controls would have been unnecessary. Physical controls are often found necessary, especially where there is not sufficient time to work out methods of creating the economic climate under which private decisions will not conflict with the interests of the community.

GOVERNMENT CONTROL AND BUILDING STANDARDS

Government control affects the standards of construction as well as the volume of work. Government building regulations are usually concerned with fire protection, public health and structural safety. They may be extended in future to energy conservation. Building regulations have usually started at the level of town governments with regulations designed to reduce the risk of fire spreading from one building to another. Regulations followed to ensure adequate drainage, proper ventilation and lighting, and structurally sound buildings.

Today, model by-laws are usually drawn up by the central government, but their interpretation and enforcement is a local matter. Clearly, they can only be drafted in accordance with current knowledge. Since their preparation tends to take a long time they are difficult to apply to innovations in materials and techniques, and tend to impede the introduction of innovations. The current tendency to draft the regulations in the form of performance standards tends to reduce the degree to which innovations are impeded, but by leaving wider scope for interpretation tends to result in less uniformity in their local application.

Governments, as with other clients, in fact more than most, lay down conditions and building standards to be met by designer and contractor. In many countries governments pay for or at least subsidise large volumes of building ordered by other agencies. For example, in Britain the government subsidises public house-building, schools and other buildings developed by local authorities. While the government leaves the local authority free to design and contract for its own building, it lays down regulations about standards and cost limits which must be met if subsidies are to be obtained. In some cases the local authority can exceed the cost limits but will only receive subsidy on that part which is within the prescribed standards and cost limits.

In addition to government regulation other bodies often regulate standards either voluntarily or with some type of statutory force. For example, in Britain and other countries following British practice, there are national standards institutes responsible for preparing and publishing voluntary national standards both relating to particular materials and to building practice. These, too, are often criticised as conservative and to favour tried materials and methods. Since these are used as a basis of specification they are said to inhibit the introduction of new materials and methods. Finance agents such as building societies, housing associations and insurance offices also impose restrictions on the use of materials and on the forms of construction which they are prepared to accept in buildings for which they will grant loans or provide insurance. Naturally they, too, tend to be conservative and to prefer the known and tried to new and untried methods. Similarly, the research stations hesitate to commend materials and methods of construction until they have accumulated adequate experimental evidence and experience of their use; this, naturally, takes a great deal of time, and in the meantime the absence of recognition naturally discourages the use of the innovation and impedes the collection of experience of its use. Moreover, the research institutes tend to report on classes of materials and not on the products of particular firms, so that the users have no guide as to which products offer the best value for money.

In some countries the designers and the contractors are legally liable for any defects which appear in the building over a period of years. For example, in France, the period of legal liability is 10 years.[99] This liability is usually covered by insurance, and insurance companies have themselves to organise testing facilities to provide a measure of the risk. In France the companies have set up building control offices, *Bureaux de Contrôle,* to satisfy themselves of the level of risks. In order to meet the problem of certifying new materials and systems of construction the *Centre Scientifique et Technique du Bâtiment* will issue *Agréments.* Laboratory tests are made and experience of site use is collected. Certificates are issued for 3 years and are renewable subject to satisfactory performance over the period. The certification may be subject to changes in the product and to limitations on its use. Thus the system provides a generally accepted level of quality based on a continuous review of the performance of the products tested. Such a system would seem to fulfil a useful function and be less conservative towards new products than individuals would need to be. In 1962 a number of countries, France, Netherlands, Portugal, Spain, Belgium and Italy, came together to form the European Union for *Agrément.*[99] The *agrément* Board in Great Britain works in association with the European Union. In Britain the National House-Builders Regestration Council inspects the work of it members and provides guarantees and insurance for ten years to people purchasing houses built by its members.

THE INFLUENCE OF TRADE UNIONS AND TRADE ASSOICATIONS

Both trade unions and trade associations are concerned with the sectional interests of their members. The pursuit of these interests will generally lead to attempts to gain advantages for their own members at the expense of other members of the community but not necessarily at the expense of the community as a whole. Trade unions can and often do follow restrictive practices which hinder the introduction of innovations and result in a waste of resources. If such measures are pushed too far they may be self-defeating and lead to a rapid adoption of new techniques which employ labour not covered by the trade unions. Short-term difficulties can be caused by strikes and other industrial protests. In the long run, measures which raise labour costs tend to encourage the use of mechanical and other industrialised methods.

Trade associations can act to encourage efficiency by encouraging better methods of operation, standardisation, and by laying down higher minimum standards of performance. Frequently, however, they are more concerned with protecting existing forms of trading. In some countries, for example, Britain, some sectors of the supply of building materials and components have been found to be heavily committed to operating restrictive trade practices.

RESEARCH AND DEVELOPMENT

Broadly research consists of studying materials, their properties and use, the way building designs can be executed and will function, operation management, costs and so on, and testing hypotheses through controlled experiments. Development on the other hand, consists in applying the research results to full-scale production of materials and buildings.

The expenditure on research and development in relation to turnover is usually taken as a rough measure of the scale of such activities. On this measure research and development in construction appears to be much more limited than in most industries. In Great Britain, where research and development expenditure is generally low in relation to other countries, about 0·5 per cent of turnover is spent on it compared with about 3 per cent in manufacturing industries and as much as 10 per cent for some industries.

Much of the construction industries R & D expenditure is paid for by the government, much of the work being carried out by the Building Research Establishment. Similar government Research Stations have been established in other Commonwealth countries and in some other countries. Such establishments concentrate on research rather than development and often mainly on government needs so that private industry and even other public authorities obtain research results as a by-product of government research demand. Generally the funding of this research is tight and many aspects of construction are not covered. Subjects once studied are often dropped even though materials, plant, methods of construction and the way buildings are used all change. This lack of continuity in research is reflected in this book which frequently has to be based on research results obtained many years ago.

Generally R & D expenditure by firms in the construction industry is very limited, especially in the case of contractors and professional firms. The structure and operation of the industry does not create conditions conductive to R & D work. Most firms are small and subject to wide fluctuations in demand. They often handle a wide and changing range of work and their problems are to an extent unique to each contract. Difficulties in patenting construction technology tends to make it difficulty to claw back the benefits of R & D from other firms. As a result such R & D as is carried out tends to be limited to problems particular to their work and organisation.

The need for more R & D is indicated by failures which have occurred where innovations have been introduced. There have been failures in technology such as unaccepted levels of dampness and noise in dwellings, in high-alumina cement technology, and in box-girder bridges. Many of these failures have been very expensive. Resources have been used uneconomically in both temporary and high buildings. Many housing schemes, particularly high flatted ones have become socially unacceptable and have been demolished. Despite the ingenuity shown in developing systems building much of it has proved uneconomic.

CHAPTER THIRTEEN

THE CONTRACTING INDUSTRY

TYPES OF FIRM

Some confusion always arises in discussions of the contracting industry as a result of its dual function as a constructor of both buildings and civil engineering works. There is really no clear dividing line between these two types of work although they can involve very different problems. On the building side the type of work ranges from rewashering taps, painting and decoration, and small repairs, through to the construction of buildings of all sizes. As the size of the building grows so the importance of craft processes tends to decline and the construction process becomes more and more a matter of fixing steel, placing concrete and earth movement, and other processes requiring the use of machinery. But while the construction of large buildings perhaps has more in common with what is generally thought to be civil engineering, the range of civil engineering jobs is nearly as great as that of building jobs. Many jobs contain elements of work both of a building and civil engineering nature and many organisations undertake both types of work. While it is possible to distinguish between specialist and general contractors, it is difficult to find any clearcut division between building and civil engineering contractors, except, perhaps, for the specialist civil engineers. In some countries there are separate bodies both on the employers and on the labour side of the industry, dealing respectively with building and civil engineering, but often a large proportion of the members belong to both types of organisation. Generally, the problems of contracting are more important than the special problems of building and civil engineering. It seems rather unprofitable to try to distinguish between the two types of contracting and here the practice usually adopted of regarding them as part of a single industry will be followed.

There are, broadly, three types of contractors: general contractors, trade contractors and specialist contractors. The general contractors usually carry out the major trades, building brickwork and masonry, excavating, concrete work and carpentry. It is these contractors who act as main contractors in those countries in which main contracting operates, and as leading contractors where trade contracting is used. The trades contractors concentrate on the work of one trade and either obtain contracts directly on their own account, or act as sub-contractors, depending on whether the practice is for trades or main contracting. The specialist contractors tend to concentrate on one particular type of work, rather than on one trade or craft. They specialise on such work as the erection of concrete frames, on electrical work, on lift installation, or on the installation of heating systems. Often such firms themselves manufacture the components which they install. Again, such firms usually act as sub-contractors or trades contractors. Local authorities, public boards and large firms often have direct labour forces to carry out their own work.

Contractors and direct-labour organisations tend to specialise by type of work as well as by trade. The smallest firms tend to be either firms working in a single trade or specialisation and acting either as trades or sub-contractors, or as jobbing builders undertaking mainly repairs and small alterations. The medium-sized firms tend to be either the larger trade and specialist firms or the general contractors working either in one town, or in a region. Frequently they specialise in house-building, shops, offices or schools, or in small civil engineering jobs. The largest firms are usually horizontal and vertical integrations, often consisting of a large number of semi-autonomous units which are sometimes organised as separate companies. Thus, often the largest firms differ from the medium-sized ones in their financial organisation, rather than in their methods of operating contracts. The largest firms have separate departments or companies to handle the various types of contracting work, for speculative building and for the production of various building components, such as precast concrete units, and to provide specialised services such as contractors plant. Direct-labour organisations are usually mainly concerned with repairs and decorations, and small extensions to the premises of the firm employing them. Central and local government direct-labour organisations are also employed in the construction of new buildings and works, particularly of housing and other small buildings.

Except in remote rural areas there are usually several contracting firms engaged on each job. The general contractor is usually responsible for the main structure, the services and finishing trades being carried out by the trade and specialist contractors. Often eight or nine firms will be employed, where the job is large, and where a great deal of specialised work is necessary the number of contractors may be two or three times as many. The number of separate contractors is not greatly affected by the system of contracting, although probably the number of contractors tends to be a little less under the main contracting system usual in England and Wales and in America than under the trades contracting systems used in Scotland and on the Continent. Where the main contracting system operates, the craft trades are sometimes let labour only, virtually a piecework system.

SIZE OF FIRMS

Often between a quarter and a third of the work of the building and civil engineering industry is repairs and maintenance, much of which consists of minor jobs to dwellings and other small buildings. This work creates a need for local contractors and much of it involves the employment of only one operative and little or no plant. The efficiency with which such work is carried out depends more on the skill and integrity of the individual operative than on the organisation which can be provided by the firm itself and, clearly, is a field in which the small firm is likely to have advantages. Much of the new work is also rather small, often single houses, shops or garages. Again, especially since several contractors may be involved, this is work for which the small firm is suitable.

In the circumstances it is not suprising that in most countries the construction industry is characterised by the small firm. In most countries something between a third and two-thirds of the firms in the industry employ less than two or three operatives.(41) Often as many as a third of the firms employ no operatives at all, the work being carried out by the partners in the firm. Recent figures show that in Britain a third of the firms employ no operatives; in France the figure is nearly a half. In Belgium two-thirds of the firms employ less than five workers; firms with five or less workers predominate in most countries, in Italy over half are in this class, in France over seven-eighths, in the Netherlands three-quarters, and in Sweden and Denmark about four-fifths. At the other end of the scale considerably less than one per cent of the firms employ as many as 500 operatives; in small countries such as Denmark there are few firms who employ as many as 100 operatives. The average size of firm is consequently very small; in France it is only three or four operatives; it is about seven in the Netherlands, ten to fifteen in Belgium, Britain and Italy and about twenty-five in Germany. The typical firm is naturally larger in countries with a large population and in which there is a larger concentration of people in large urban areas consuming a large proportion of the national product in the form of the products of the construction industry.

Direct labour organisations also vary in size depending on the size of their parent organisation and on the degree in which it is concentrated into one district, and, of course, on the amount of work the organisation is called upon to carry out.

ECONOMICS OF SIZE

There is no general optimum size of firm or organisation, the best size depending on the nature of the work, the conditions under which it needs to be carried out, and the nature of the organisation and the ability of the management. As already mentioned, much of the maintenance work and much of the constructional work to small buildings requires so few operatives at a time that the large firm cannot offer any advantages over the smaller firm. As jobs grow in size more labour is needed and more plant and organisational ability. Much of the plant and the managerial services can only be obtained in indivisible units which are sufficient to meet the needs of fairly large labour forces. For example, a crane will need a succession of fairly large jobs to keep it utilised; similarly, a programmer will be able to handle a large number of jobs each year. Thus, specialised equipment and services can only be utilised efficiently providing that the firm is of a certain minimum size. The larger sized firm is also in a better position to keep its operatives fully occupied, especially the more specialised gangs.

But the larger the firm the greater the need for co-ordination. Where the firm is so small that the principal works on the job, it is fairly easy to co-ordinate the work of the firm. In such circumstances the amount of equipment and staff is small and their use can be appreciated visually without the need for elaborate records. Once staff and equipment are spread over a number of jobs problems of allocations arise and records must be kept of where they are, what they are doing, and when they will be free for use elsewhere. Paid supervisors must be employed and incentives provided to encourage a high level of output and a proper standard of work. If the labour and other resources of the firm are not to be left idle new work must be obtained and dovetailed to fit in with resources as they become available. It must be possible to obtain sufficient finance to meet the standing charges of labour and equipment.

While the problems of co-ordination increase with the size of the firm, size does, of course, bring many advantages. Large firms can usually obtain finance more easily than small firms and at cheaper rates. The price paid for materials usually depends on the size of the order, the larger the order the lower the price. While smaller firms can increase the size of their orders by purchasing materials ahead of needs, this raises costs of handling and storage, and of interest on the capital employed. Price reductions can usually be obtained by purchasing equipment on a larger scale. The greater the amount of work, and the spread and size of jobs, the greater the opportunity of keeping the resources in use and of obtaining economies from specialisation. Larger firms can usually obtain better staff and can afford to develop better techniques than small firms. The larger the amount of business, the easier it is to spread the risk of particularly unfavourable contracts. With better access to credit and a larger demand for materials large firms can more easily set up units to produce the materials they require, or use their large purchasing power to secure more favourable terms from manufacturers than can small firms. Moreover, large firms can set up their own selling organisations and secure control over their market.

However, the greater the size of firm, the more complex the organisation, the longer the chains of command and the more remote the control and the level of incentive on the job. As size increases more non-productive workers need to be employed to co-ordinate the organisation. Their cost must be balanced against any savings they achieve in the productive labour force or in the economies of scale. While size may enable the risks to be spread, size does not necessarily increase profits; it may result only in evening out the spread of profits and losses. Moreover, the capacity of individuals to manage is limited and firms and organisations can outgrow the capacity of their managers. In many large organisations considerable local autonomy is given either to reduce the scale of management at the top and to provide for greater flexibility, or because of the importance of local factors in making decisions. In fact, some large firms consist of several medium-sized, largely autonomous firms, linked only for the supply of capital and other central services.

Statistical evidence of the economics of scale in construction work is limited. It is difficult to measure construction ouput and costs, and labour expenditure is often not recorded either precisely or completely. The variability of contracts and work creates comparison difficulties. Most statistical studies have been confined to house-building, for which the difficulties are less than for other types of building. Studies of house-building in the United Kingdom have provided some evidence that returns increase with scale from the smallest to those slightly larger and clear evidence of decreasing returns over the middle range of contract size. The evidence suggests that house prices tend to fall, other things being equal, as the size of firm increases from very small to somewhat larger but to rise as firm size increases further. A similar relation was found between labour productivity and size of contract. Since large contracts can only be tackled by large firms, this suggests that where possible contracts might profitably be split between a number of small contractors.

There is, in fact, no ideal size; the most economic size depends on the conditions of the industry in the country in question, on the type of product or service required, and on the ability of the principals. An industry is not, therefore, necessarily inefficient because it contains a large number of small units; this may be the best arrangement in the circumstances. While the increase in mechanisation, in the development of management techniques, in the size of jobs, and in the development of system building all tend to increase the size of firms likely to provide the highest level of efficiency, smaller firms can achieve comparable efficiency if specialist firms exist who can suppy the plant and management expertise required.

THE NATURE OF THE INDUSTRY

The construction industry differs from many other industries in a number of ways. It is largely an assembly industry, assembling on site the products of other industries. Because the products of the construction industry are designed to meet the needs of a large range of users, the assembly process tends to require a considerable amount of knowledge and skill. The scale of buildings and works and the large range of processes involved has led to the practice of providing only a general guide to the designers' intentions in the plans, the operatives being expected to have the knowledge and skill needed to assemble the components. The need for this sort of knowledge and skill has probably limited the possibilities of devolution. As already indicated, the range of products is perhaps greater than in most other industries, and while specialisation amongst the firms in the industry reduces the range of jobs tackled by most firms, the range is still probably larger than in most other industries and to a large extent each job is unique.

Construction work is largely site work and is hence subject to the hazards of working in the open air and on disturbed ground. It is often difficult to provide much protection from the weather, at least in the early stages of construction. In some countries earth movement is quite impossible and concreting and mortar work is impeded during the winter. In recent years the development of special plant and protection has reduced these types of hazard. In some countries output is reduced either as a result of the heat or because of the unpleasantness of working in the heat or wet. The exposure to the atmosphere and to site conditions tends to reduce the life of the plant employed in relation to the work it carries out. As explained earlier, it is difficult to utilise plant on site as fully as in a factory.

A contracting industry poses problems of organisation additional to those arising where goods are made for the market. While meeting the needs of the wholesale and retail markets poses problems in selling, once these are solved the requirements of the market are usually fairly predictable and regular, at least in the short run. The range of products and the output required can be determined and their production planned in detail for some time ahead. The contractor has far less opportunity of determining his future market. The probability of obtaining any particular contract is often only about one in ten and in some cases much less. It is, therefore, necessary to bid for many more contracts than required with the inevitable risk that at times the number of contracts secured will be far less and at other times far more than can be handled efficiently by the organisation. Contracts must usually be started a short time after they are awarded, so that a contractor rarely knows even a short time ahead upon what contracts he will be engaged. Under such conditions it is difficult to plan the future load of work so as to secure the most efficient use of the firm's resources and wastage is inevitable. Since the range of work is wide and uncertain, and since, especially for large firms, work is widely distributed geographically, much of the labour in the industry is casual. Generally, casual labour is less efficient and has less incentive to be efficient than labour on long-term engagements.

The two major elements of building costs are labour and materials. Their proporitons depend on the type of building and the relative local levels of prices. In Britain just under a half of the value of building work is represented by the costs of materials; of the balance two-thirds to three-quarters is the cost of labour. The proportion for materials is higher for new work than for repairs and maintenance: for housing, materials account for a half to two-thirds of the costs, about a third being the costs of labour. In the case of housing, overheads and profits generally amount to about 10 per cent of the total costs. The ratios depend very much on the definitions used. The ratio of material to labour costs naturally rises with increases in the prefabrication of the components. This ratio is generally lower for civil engineering work where the materials are mainly very simple and involve little preparation. Again, in some countries labour is relatively expensive as compared with materials. For example in America the materials to labour ratio was found to be about 43 to 57, [16, 100] in spite of the added incentive to prefabrication where labour is expensive. Generally, the ratio of overheads and profits in the construction industry is fairly low, perhaps something of the order of a fifth to a sixth. This is probably partly a result of a low use of management skills, partly because of the absence of large selling organisations, and partly because of low capitalisation.

CAPITAL NEEDS IN THE CONTRACTING INDUSTRY

The comparative smallness of the capital needs and the comparatively high ratio of liquid to fixed capital tend to result in the industry being easy to enter. Ease of entry may lead to the entry of firms without adequate experience and capital to sustain the development of contracts let alone expansion, and perhaps be the reason for the comparatively high wastage rates in the industry. While no very reliable figures are available, it would appear that total capital is about equal to the annual wages of the operatives using it, and that only about a half is fixed capital. Stores and workshops on an extensive scale are not generally required. The needs for plant are comparatively small and much of this can be hired. Materials are mainly ordered to meet current needs and usually these can be purchased on credit sufficiently long to cover the period until the cost of the materials is recovered in an interim payment. The interim payments also reduce the period over which labour needs to be financed. The reliance on borrowing liquid capital tends to make firms vulnerable to changes in the level of receipts and payments, and to financial conditions. Up to a point, additional work can be financed by using the credit provided by the materials merchants. There must, however, be some lag between expenditure and receipts. As long as the firm is operating at an even level the receipts from previous jobs will normally be sufficient to meet the temporary difference between payments for labour and materials and interim payments on current work. Difficulties arise where the rate of expansion is too fast, or where general financial difficulties lead to the withdrawal of credit facilities.

The intensity with which assets are used and the margins of profit vary considerably.[(101)] The effect on the return on capital a proper use of assets can be great as of a proper use of labour on the site.[(101, 102)] A small sample of firms in Britain showed operating profits of from 6 to 40 per cent.[(101)] Good rates of return were obtained from comparatively poor operating profits where the assets were used intensively and material stocks and work in progress were kept low.

Overheads and profits are affected not only by management's use of their resources in the short and long run but by how they handle changes in demand for their services as workloads decline, overheads per unit of work rise and profits can change substantially in relation to turnover and capital.[(103)]

CHAPTER FOURTEEN

THE ORGANISATION AND OPERATION OF THE BUILDING FIRM

TYPES OF ORGANISATION

There are many forms of organisation employed in building firms, the form depending on the size of the firm and the nature of the work it normally undertakes. The principals of the smallest firms usually work with the operatives at one of the crafts. In firms a little larger the principal is more likely to supervise than to work at a craft, but at the same time will perform such functions as estimating, programming and purchasing materials. The principals of larger firms will usually concentrate yet more on the more purely managerial functions, leaving site supervision and day-to-day control to paid foremen. The organisations amongst the large firms tend to vary considerably from one firm to another. In really large firms the principals are mainly concerned with policy and with top-level administration, all the execution being left to paid employees. In some cases firms are divided into a series of separate divisions which are almost autonomous, the divisions being partly functional and partly geographical. The functional divisions are particularly appropriate where such diverse work is accepted as house-building and civil engineering, or, as frequently happens with general contracting firms, where the firm has stepped outside its normal function of construction work and has set up trading and manufacturing departments, such as joinery shops, quarries and plant-hire departments. The regional divisions follow from the regional nature of much of building organisation, the need for local labour, materials and plant supply. In such cases the regional divisional headquarters control the contracts, the responsibilities of the site organisations being limited to day-to-day control. In contrast, some firms concentrate a large portion of their organisation on the sites which are almost autonomous, making most of the decisions and accepting responsibility for accounting, ordering materials and plant, labour control and other contract matters. Where this arrangement is adopted the activities of the head office are confined to overall administration, tendering and finance. It seems doubtful whether sites can be organised with this degree of autonomy unless they are very large.

As mentioned earlier, managerial problems are to a large extent a product of size. The small trade contractor with only one or two employees working beside their principal has few managerial problems beyond obtaining a continuous flow of work and purchasing a few materials. The small firm undertaking small housing contracts has, similarly, few managerial problems since the principal is himself in full control of every element of the contract. The problems increase in magnitude with the growth of size and diversity of functions. Broadly, two kinds of problems arise; the problems of ensuring that all the units are properly co-ordinated and correctly balanced to fulfil the functions of the firm, and the problem of ensuring that each unit of the firm functions economically.

Normally the central activity of a building firm is obtaining and carrying out contracts for building works. All other activities are centred around these central functions. Many large firms have expanded horizontally into the many different types of building work, regionally to cover a whole country and often countries abroad, and vertically to production of building materials and to plant hire, and at the market end into building speculatively and to the development of property.

SECURING CONTRACTS

At the centre of contracting is the problem of obtaining and choosing new business. In the short run the firm has a given set of capital assets and a staff of executives, technicians, clerical workers and an operative force. The aim is to employ these resources profitably, which at least means covering all costs, including overheads of all types and obtaining a normal return on the capital employed. Firms may be more ambitious and wish to obtain the highest possible profits; this will usually entail maximising the turnover of the assets. In the long run ambitious firms will wish to expand; more capital will be needed. Capital may be obtained by ploughing earnings back into the firm, borrowing, or by issuing shares.

In the short run, overheads are fixed. Operative labour can, of course, be discharged if no work is available but usually a firm will wish to retain at least its key operatives, and in times of full employment other members of the labour force, because of the difficulties of recruitment once labour is lost. "Overhead" is a generic term which applies to costs which cannot be related to some particular item of work and which during the period considered cannot be reduced by making changes or reducing the work load. For example, it is not possible to dispense with a site foreman, or to reduce his cost by marginally reducing the operations carried out on the site. Similarly, it is not possible to dispense with a regional manager because there is one job less to supervise in his area; nor is it possible to close a head office building when the number of contracts secured falls marginally.

The length of time for which overheads are fixed depends on the way in which they arise. Site overheads are normally fixed for the duration of the contract; other types of overheads are fixed until the load of work is substantially changed and until the organisation can be adapted to a new situation. In the long run, staff can be dismissed, capital assets sold and hired goods and credit returned to the lender. Such changes take time; such factors of production can only be reduced or increased gradually. In the short run these indirect factors of production are fixed; if the volume of work falls their use will be spread over a reduced volume of work and the percentage for overheads will rise; if the volume of work increases these assets and staff will be spread more thinly and the overhead percentage will fall, but, of course, efficiency may suffer.

Since, in the short run, the factors which give rise to overheads are fixed, there is no saving if less business is handled. Overheads are fixed and hence any work at a price which just more than covers its prime costs is worth accepting since it makes some contribution to the overheads. Hence it pays any sort of organisation, whether a profit-making organisation, or a direct-labour organisation, to take on a job as long as the price covers the costs which directly arise from the job itself, such as operative labour, materials and plant hired in for the job, and which makes some contribution to the general overheads.

Clearly, in the long run contractors wish to cover their overheads and to earn a reasonable return on their capital, but in the short run it may pay to accept contracts at prices which only just more than cover their prime costs. The margin to add to the estimated prime costs of a job depends in the main on the balance currently achieved by the firm between the load of work on its order book and its capacity, on the balance between fixed overheads and the excess over prime costs provided by current contracts, and on the state of the market and in particular on expectations about the behaviour of rival contractors. Strategies for tendering have been considered.[104] At times when there is little work being put out to contract, a firm with a great deal of spare capacity will accept contracts at prices which only little more than cover the prime costs of the job. In contrast, a firm with a full order book and with some indifference about obtaining a particular contract will put in a tender price which provides a generous return on the capital employed as well as covering a proportionate share of the overheads.

The determination of the right level of mark-up to add to the direct costs calls for considerable judgement of the market and needs to be based on some firm information on the short-term position of the firm and on the conditions in the market. The short-term position of the firm is assessed by comparing the sum of the overheads which have risen over the period of account with the contribution to the overheads obtained from work done during that period. The forward position is obtained by allowing for the contribution expected from current contracts. Some allowance for normal profits on the capital employed can be included in the overheads or it can be treated as a separate item. The position in the market can, of course, be judged to some extent by the results of previous tendering. In some countries the results of public tendering are published and provide a guide to market conditions. Further information is published in some countries of the value of work coming on to the market. This takes various forms; sometimes information is available of the value of work done, or better the value of work for which contracts have been signed, forward indicators are sometimes available in the form of commissions to architects for designs, or to quantity surveyors for the preparation of bills of quantities.

As mentioned earlier, materials usually account for something just under half the costs of the job; of the remainder, direct labour accounts for two-thirds to three-quarters, the balance being overheads and profits. A reduction in the amount of contracts much below the level the firm is geared to handle, or losses on contracts, soon upsets the balance between overhead expenditure and the margins earned to cover them. Very high mark-ups are then necessary to restore the balance and these can only normally be obtained during periods when the industry is overloaded. It is not suprising, therefore, that contractors place a great deal of importance on maintaining a full load of work. Usually a building owner will invite at least six contractors to tender; often the contract is advertised and any number of contractors may tender. Thus the chance of securing any particular tender is often quite a small one and contractors need to tender for many contracts in order to secure an adequate volume of work. Unless the success rate remains fairly constant this tends to result in some contractors getting overloaded with work and others having insufficient and tendering on an even larger scale in the future. Moreover, clients usually expect work to start on the site within a week or two of the contract being signed, so that where a lot of work has been secured contractors tend to have far too many contracts over which to spread their resources, the contracts take too long and clients are dissatisfied; overheads tend to rise because indivisible units of production are insufficiently used.

Tender prices submitted by various contractors for the same job tend to vary considerably. A very high tender is, of course, often only a cover price put in by a contractor who does not require a particular job but who does not wish to appear to decline to quote. Variations in tender prices also reflect the extent to which contractors already have sufficient work and the lower price at which some contractors can tender either because of above average efficiency, or because of special advantages for a particular job, perhaps because of the location or their experience, or perhaps because they reduce the use of materials and labour to the minimum that the client and his consultants can be induced to accept. Variations also arise from errors in estimating and in fixing the right level of mark-up over the net cost.

The estimating of the cost of carrying out a particular contract is very hazardous and it is generally believed that the margin of error is often very wide. While this may lead to both the loss of particular contracts and to some contracts being taken at prices far too low and others at generous margins, the uncertainty may result in price levels, in general, being raised to provide an allowance against such uncertainty rather than in the reduction of the general profit levels of contractors. Another result would, of course, be to turn building into something of a gamble for clients, since it might be more or less chance whether individual clients obtained prices above or below average.

Since under competitive tendering it is normal for the contract to be awarded to the contractor with the lowest tender, there is a tendency for the contractor who underestimates the cost to obtain the job. In the long run the tendency would be for contracts to be obtained at low or no margins and the contractors would tend to achieve profit levels below those considered reasonable in the light of profit margins in other industries and comparable conditions. The general level of mark-up on the direct cost would therefore tend to be raised so as to restore normal profits. Thus underestimating the direct costs might only result, in the long run, in higher mark-ups in order to restore normal profit margins.

The difficulty of costing and the danger of tenders being accepted which were based on an underestimate of the costs encourages contractors to cover themselves by adding an allowance for contingencies. The presence of contingency allowances tends to reduce the incentive towards efficiency and so to lead to still higher contingency allowances. Thus costing difficulties may tend to lead not only to unfairness between clients but to higher prices and to lower efficiency in the industry.

COSTING

The estimating of costs is of particular difficulty in any form of contracting industry, especially in one, such as the construction industry, where the products are so heterogeneous and the cost dependent on many factors outside the control of the contractor and are almost impossible to forecast.

In some countries, notably Britain and other countries following British methods, contractors are normally supplied with a bill of quantities which sets out all the materials and labours necessary for the construction. In other countries the contractor is responsible for taking off his own quantities, these being determined from the plans and specifications. The work can then be costed by applying unit prices to the quantities. Reference books are available which list for each unit of work the quantities of materials required, the allowance for waste, the requirements for skilled and unskilled labour, and the requirements for plant. The quantities can then be costed at current prices. Standard allowances are often given for supervision and other overheads. But such figures are inevitably only averages; their value, apart, perhaps , as a basis for measuring and costing for materials, is often questioned. The actual labour and plant times will vary from one contractor to another and from one site to another in accordance with the efficiency of the firm and its labour, the exact nature of the work, and local site factors.

Firms usually compile their own unit rates based on their own experience. While lists of such rates may not be as comprehensive as those given in the reference books they do relate more closely to the conditions under which the firm works and at least provide a link between the experience of the firm and the reference books. But it is difficult to list rates suitable for every unit of work under all the conditions in which they might be performed. For example, the labour and plant per cubic yard of concrete depends not merely on the type of mix but also on such factors as the degree of mix accuracy required, the shape of the formwork, the amount of reinforcement to be used, the rate and period over which the concrete is required, and the total amount. The use of unit rates for estimating is often criticised and alternative methods have been developed.

The unit-rate method is probably accurate as far as materials are concerned, since, in general, the materials requirement is related to the building to be provided and not to the method of erection but wastage rates may be variable. The requirements for labour and plant are, however, largely a factor of time and this depends on the exact nature of the construction process and the way it is related to all the other activities. In order to estimate how much labour and plant is required it is necessary to visualise how the work will proceed and how each operation will be organised. As mentioned earlier, an operation is taken to be a piece of continuous work. The method of estimating the costs of the direct work usually advocated is to base the estimate on the operational programme. Under this system the first step is to work out the programme of site operations, and for this purpose the critical path method of analysis is now available, and then to add up the operative and plant days necessary for the process. The requirements for materials can be estimated in the traditional way or from the operations; the latter method carries the advantage that the material requirements can then be listed according to the date they are required. The disadvantages of the programme and production-plan method of tendering lie in the time taken to work out the programme in the first place, which is wasted if the contract is not obtained, and the frequent difficulty in obtaining adequate plans in time to prepare the programme. Difficulties in applying this method also arise because of the short period often allowed for tendering.

DATA COLLECTION

Unit rates become out of date and are only strictly applicable to the organisation for which they were observed. Regular measurement of operations either by site observation or by the analysis of recorded man-hours is necessary to establish and maintain a sound basis of estimate. Usually the contract cost data are not generally suitable for costing purposes.[105] A better basis is provided by the measurements of work done for the purpose of interim payments, together with time sheets. The records maintained in connection with bonus schemes often provide suitable data, since the bonus targets are often set in terms of homogeneous units of work or in terms of operations. Site observations must be repeated several times over if reliable results are to be obtained. This work is often combined with method study and with other operational studies aimed at improving output. Such work lies outside the conventional work of estimating departments and usually requires the setting up of a work study and statistical department. The creation of these departments as an aid both to estimating and to the improvement of the efficiency of operations is another step in the rationalisation or industrialisation of the construction industry.

OTHER CENTRAL SERVICE DEPARTMENTS

The estimating and work-study departments are just two of the many central service departments which contractors need. There is, of course, not necessarily a separate department to handle each function; one department will often handle several functions, especially in small firms. The number of central service functions tends to grow as more and more attention is paid to securing high levels of efficiency and management becomes more scientific. The use of industrialised methods creates a need for a more scientific attitude to the management of contracts. Programming and bonusing functions are now frequently added to the more conventional functions of buying and surveying. Frequently, too, designers are needed to design special components and form work. In addition to the special functions associated with contracting, the normal functions of business management must also be carried out and staff are necessary for secretarial, legal, personnel and accounting work, and often for market research and estates management. The way these and other serices are provided and organised varies considerably from one contractor to another.

OPERATIVE AND NON-OPERATIVE DEPARTMENTS

In addition to the operative and service departments, contracting organisations frequently have trading and manufacturing departments, which might be called non-operative departments. Such departments arise as a result of contractors trying to increase their control over their supplies by setting up their own supply departments, or by trying to secure control over sales by setting up estates departments to manage speculative developments. While such departments are usually set up to supply or to handle the products of the operative departments (the contracting departments), non-operative departments, once set up, frequently develop interests of their own, provide services direct to the market and function as independent firms; often such independence is the only way in which efficiency can be secured. In this way contracting firms often enter such fields as plant hire, quarrying, joinery manufacture and the manufacture of precast concrete goods. Frequently the amount of business with other firms exceed that with the parent firm and the department is split off and becomes a separate company. Some contracting firms become speculative developers and property companies on a significant scale.

There is a tendency for management to give far less attention to trading departments than is warranted by their often considerable size. There is a tendency to look upon trading departments as subsidiary units existing only to serve the interests of the contracting departments. Joinery and plant departments are often considered necessary if certain types of contract are to be secured; their efficiency is often thought to be of secondary importance and their plant is often seriously underemployed. Similarly, the benefits of direct buying are often overestimated, and the possible savings from cutting out the middle-man are accepted without setting against them the handling and stocking charges so incurred. This attitude often results in the acceptance of lower returns on the capital used in the trading departments than on capital used in the contracting departments. Generally, the non-operative departments are only likely to be run efficiently if the criteria of the success of the management is in terms of the return on the capital employed.[106] This does not prevent discounts being given to other departments within the firm. It is when the trading departments are run entirely for the convenience of other departments that inefficiencies are likely to arise.

CONTRACTORS' PLANT

A good example of the contradictions between the interests of various parts of the firm which can arise in connection with the non-operative services is provided by the plant departments. A plant pool is perhaps the most usual trading service which contractors provide for themselves. There is a tendency amongst contractors of all sizes to try to be self-sufficient both for mechanical and for static plant. As mentioned earlier, the difficulty is to keep the plant in constant use. The problem falls into two parts; the first is for the plant pool to keep the plant turning over rapidly, and the second for the site organisation to achieve a high degree of utilisation on the site.

Hire charges can only be kept at a low level if demand and availability are closely matched, although, of course, some excess availability is necessary in order to provide flexibility, and because specialised plant not usually available from plant-hire firms may have to be carried even although it is not often used. The greater the degree of self-sufficiency the greater the risk of excess capacity during periods of slack business; if these occur during general slackness in the industry it may not be possible to hire the plant to other firms. Such risks can be reduced by the limitation of the fleet to that necessary to meet normal needs. The prices which are important to the contracts are the delivered prices; where the contracts are at a great distance from the plant depots the delivered prices may be greater than the prices charged by nearby plant-hire firms. Firms whose work is geographically widespread may not find it economical to own much plant, especially if the contracts are of a short duration.

Where the demand for a particular type of plant is large it should be possible to keep the time the plant is idle in the yard to no more than is sufficient to enable it to be properly maintained, so that site breakdowns can be eliminated. Only the largest firms have demands which satisfy these conditions, but even in these cases the fleet is often too large and the average periods of idleness are often excessive. A solution sometimes put forward is that firms in the same area should combine to form one large pool, when there would be a greater chance that fluctuations in the levels of demand would cancel out. While such a central reserve might result in a greater utilisation of the plant, there is some doubt whether a jointly operated pool would operate smoothly, and it is possible that an independently owned pool, that is a plant-hire firm or a co-operative, is the more satisfactory solution. As operated in the present environment, it is doubtful if the plant-hire firms obtain the maximum utilisation. Since contractors usually own much of the plant they need, the plant-hire firm tends in part to be a reserve of last resort. It is notable in some countries that the plant-hire firm charges relative to the contractors true costs are greater for the types of plant normally owned by contractors than for types more normally hired.

Where there is a plant pool there may be a conflict between the separate interests of the pool and the contracts. On the one hand, the pool can most easily balance its revenue and expenditure by limiting the plant fleet to those items in high demand and by making high charges for the plant. On the other hand, the contracts can reduce their effective costs not only by intensive usage but also by loading beyond the stated capacities of the machines, and by reducing the time spent on cleaning and day-to-day maintenance. The method of charging must, therefore, provide incentives to both sides and encourage methods of operation which are in the common interest, that is, in the interest of the firm or organisation as a whole.

Plant charges can take one of two forms, either a percentage oncost on the value of the contract, or a charge for each item of plant for each unit of time it is borrowed. The attraction of the oncost method lies in its simplicity, since little in the way of records or costing is necessary. But the contracts bear a charge unrelated to the services obtained and there is no incentive to careful or intensive usage, or discouragement to the wasteful accumulation of plant on the site. The oncost charge method provides no figures on which to base comparative assessments of alternative methods of carrying out the operations, no measure of the efficiency with which the pool is operated, or any check on the size or balance of the fleet. Percentage oncosts have been found unsatisfactory both for mechanical plant and for small tools and static plant; in fact, it appears that when used for static plant and tools a percentage oncost is very likely to lead to their being treated as consumable stores.

The inventory system is often adequate for plant such as small tools and static plant which are of insufficient value to bear appreciable administrative costs and which require little in the way of maintenance. This system consists of comparing the valuation of the items of plant prior to going on a contract and on their return to the depot. A usual arrangement is for the contract to be charged on the basis of the inventory as the plant is supplied, each item being valued as unused, 50 per cent used, and so on. The inventory is revalued at the completion of the contract, items badly damaged or missing being graded at zero per cent; the contract is credited on the basis of the revaluation.

For large, more valuable plant the inventory system is not usually adequate and a straight charge for each working day appears to be the most satisfactory of the alternatives. It is simple to calculate and provides a greater incentive to planned machine use than does a charge levied only on the actual machine use. The charging of a lower rate for idle time complicates the calculation of machine charges, reduces the incentive to intensive use and is often open to abuse. Safeguards to ensure fair prices to the contracts and to provide incentives to the plant pool can be provided by charges ceasing if the plant breaks down, by the requirement that it is replaced without extra charge, and by requiring that the pool must supply plant required at the standard rates even if it has to borrow the plant from other firms.

Some contractors, chiefly those operating large projects with lengthy contract periods, purchase plant especially for each job and dispose of it at the end of the contract. Each contract is usually responsible for servicing its own plant. Those who operate in this way claim several advantages for the method. The undertaking is relieved of the problems of central plant administration, and plant can be chosen specifically to meet particular requirements, whereas where there is a plant pool a less satisfactory piece of plant may have to be used because it is available. The costs of haulage, possibly from a distant depot to the site, is avoided. It is, however, unlikely that the savings in cost are as considerable as is often suggested. The reserve of plant on the sites may be no smaller collectively than the reserve that would carried by the plant pool. On the other hand, more machine tools may be needed for use in the various site workshops than in a central fitters' shop. Moreover, the economies of scale may be lost both in the use of plant and in maintaining it. The choices of plant and the prices to be paid and received will depend on market conditions and, as a result, the availability of plant will be limited at times of shortages, while losses may result in sales during periods of business recession. In addition, the fluctuations in prices make it difficult to forecast plant costs with any certainty.

The basic element in the hire charge is, of course, the rent of the plant, that is, the annual cost of replacement, interest on the capital tied up in the plant, the costs of maintenance, management costs for administration, and normal profit on the capital. On average over the years the rent has to be sufficient to meet current expenses and to provide a reserve sufficient to enable the real value of the fleet to be maintained.[68] The correct level of rent is difficult to estimate unless adequate records of the operation of the fleet are maintained. Unsually only the largest firms have fleets of sufficient size to provide operational data on a sufficient scale to yield reliable estimates of the rent elements in the annual costs. The machinery for the collection and pooling of operational data is usually inadequate. Often the elements of rents are estimated on the basis of accountants' rates of depreciation, on notional rates of maintenance, and no allowance is made for interest on the capital value of the fleet or for changes in the value of money. As a result the charges for plant bear little relation to the true costs. If the charges for plant are fixed below the true costs there will be an encouragement to use plant in situations for which it is not economic, while if changes are fixed above the true cost the use of plant will be discouraged where it could be economically used. This is a situation where taking a conservative view and fixing costs so as to recover the capital outlay in a few years can lead to the real costs being higher than necessary, since some economic uses of the plant will appear uneconomic and the use of plant will be less than it might be.

There is, of course, a choice of fuels for powering plant, the most important being petrol, diesel and electricity. Generally, diesel plant is more expensive to purchase than the other types but is cheaper to maintain than petrol-driven plant and uses a lower volume of fuel. The balance of advantage usually depends on the price of the fuel. Where diesel is taxed at substantially lower rates than petrol, diesel plant usually has lower total costs than petrol-driven plant. It is not possible to generalise about the relative cost of using electrical plant since the total effective cost of electricity is so very variable, depending on the price of the power per unit, on the costs of taking electrical mains to the site and wiring up the site. Generally speaking, the greater the use of electricity on the site the lower the cost. The effective price of electricity is naturally usually lower on sites in built-up areas than on rural sites, and particualrly where the supply is already available on the site.(69) Electrical plant will have a greater chance of competing against other types of fuel in countries where cheap power is available.

While operating costs increase with the size of the plant they do not increase proportionately if the full capacity of the plant can be used. The operating costs fall the more usage the plant provides. Thus the operating costs of plant fall with increases in site time, in nominal usage time, and in the degree of capacity used. These factors need to be examined together. Up to a point it will pay to accept low utilisation of capacity in order to obtain higher annual operating times. Too much attention to obtaining the optimum size of the plant for the job can lead to stocking a large range of plant which is comparatively rarely used. Greater flexibility can be obtained if too much attention is not given to optimum use, the same machine being used to obtain different outputs and to provide a range of services. The break-even point between optimising the machine choice for the job, which reduces prime costs, and optimising it for the organisation as a whole, which reduces overheads, can readily be found for the conditions operating in any organisation. Examples have been worked out for British conditions,(73) which indicate the wide range of plant sizes for which operating costs, that is total costs, are the same.

It would, therefore, appear that plant operation costs per unit of output can be reduced by standardisation on one or two sizes of each type of plant, rather than providing a large range of sizes. The pattern of sizes which will give the optimum results will, of course, depend on the range of work obtained by the organisation concerned. This policy will tend to result in savings not only because of the greater use of the stock of plant but also because transport costs will tend to fall and standardisation will lead to lower maintenance costs and to the advantages of bulk purchase.

Plant hire, whether from specialists firms or from co-operatives, offers a potential alternative to ownership. Clearly, plant hire is only likely to be able to offer a more economic alternative if this leads to a more intensive use of the plant available. Where the average rate at which plant is used is as low as is the case, for example, in Britain, there is scope for improvement from another method of plant holding. Clearly, plant-hire organisations would need to carry sufficient plant to satisfy all the needs that might be put upon them but, even so, the total amount of plant needed should be less than would be needed where each firm held its own plant, since there would inevitably be some pooling effect. The economies of scale would also tend to be greater since plant-hire organisations could be much larger than the plant pools of separate contractors.

In some countries, of course, for example, Britain, plant-hire firms already play a large part, but the usage they obtain from the plant is lower than might be possible under conditions in which all plant was hired from commercial or co-operative plant-hire organisations.

Contractors can, up to a point, afford to pay more per day for plant which is hired than the annual daily charge for plant owned by them. This situation arises where the plant is wanted for only a short period during the year. The shorter the period for which the particular item of plant is to be used during the year, the greater the outside hire charge it is worth incurring, or, to look at it another way, the greater the advantage to hire from a central plant pool. Methods have been developed for programming plant needs so that the combination can be worked out of owning and hiring plant which gives the lowest combined costs.(73)

MANUFACTURING DEPARTMENTS

The range of manufacturing departments maintained by contractors is mainly centred round three materials, wood, stone and concrete. The joinery and masonry works tend often to be prestige departments. Often the cost of the product is greater than it would be if purchased outside. Managements often report that such works run at a loss but are essential in order to obtain certain classes of business. It is doubtful how far this is true. There is some suspicion that the supposed necessity of the contractors joinery works is an anachronism from times before the introduction of modern high-speed joinery machinery and the growth of the joinery industry.

On the other hand, the manufacture of precast concrete is a modern development, often brought about by the development of off-site prefabrication. It usually involves little plant and may reduce transport costs. Even so, it is doubtful whether many contractors maintain sufficiently accurate records to determine whether or not their concrete products are competitive with the market.

SYSTEM BUILDING

System building often involves the setting up of a manufacturing department but it also involves much more besides. The contractor will need to set up a manufacturing department only if he uses a system that requires factory-made building units. Even so, these may be purchased from another contractor, perhaps from the developer of the system. Often the manufacture and erection of the units is an integrated process and stock piling would be uneconomic. In such cases orders to the factory depend on the rate at which contracts for the supply of buildings are obtained and it is likely that the factory will be far less intensively used or in as regular production as a normal factory manufacturing goods for direct sale to the market. As a result, capital intensive methods may not be economic; it may pay better to use methods which require less capital than is worthwhile where continuous full production is possible, because labour is more flexible than is capital tied up in plant. Except where there is some manufacture for general sales, as is possible with open systems, manufacutre will be geared to erection on the site and the manufacture of the units is really only an off-site part of the erection process.

SPECULATIVE DEVELOPMENT AND PROPERTY OWNERSHIP

Many building firms, particularly those specialising in house-building, largely develop speculatively. The speculative house-builder purchases land, obtains planning permission and any other consents, lays out and develops the land and erects the houses; these are generally sold freehold. Leasehold sales usually only occur where the contractor only owns the land leasehold. The contractor thus develops and sells houses in a similar way to most manufacturers, that is he produces for a market and builds for a customer still to be found. The price he fixes in relation to general market price levels amongst competing houses.

The extent to which builders can develop speculatively for sale depends on the financial institutions in the country in which he works. Since few house purchasers, particularly first-time buyers for owner-occupation, have the capital to purchase a house, the market depends on financial institutions, such as building societies, banks and the credit houses, having sufficient capital to lend would-be purchasers, as well as on sufficient credit-worthy purchasers. The availability of capital for mortages depends partly on the rate at which past loans are repaid and partly on the money lent to the credit houses. The latter depends on the rate of interest available from rival financial houses.

The availability of would-be purchasers to borrow sufficient to purchase a house depends on prices, the rate of interest, their available savings and their incomes. Often there is a lack of balance between available capital for lending and the demand by credit-worthy borrowers. This tends to result in a lack of balance between the supply and demand of houses for sale (all hosues, not just those currently being built) and considerable price fluctuation.

When the demand for new houses is high in relation to supply, prices rise, builders developing speculative housing achieve larger profits and bid-up the price for land suitable for building. If demand is low in relation to supply, prices fall, profits fall, builders reduce development, the price of land falls and frequently many builders go out of business.

A somewhat similar situation can exist for the speculative development of industrial and commercial building. Often, however, such buildings are not developed speculatively for sale to owner-occupiers but for sale to investment houses or for leasing. If the builder is developing for leasing, he inevitably builds-up a portfolio of properties and needs a permanent and growing supply of capital to finance the buildings. Such capital must usually either be raised by selling equity in the firm, by borrowing on the security of the buildings and their expected rent income, or by sale and lease-back.

The success of speculative building for leasing, especially of commercial property, depends less on the efficiency of the firm as builders as on their ability as speculative developers and on the financial situation in which they operate. Land and overheads tend to cost from as much to several times as much as the costs of construction. The value of land varies with the valuation of the completed properties, which in turn depends on expectations about future rents. This depends on expectations about physical demand and general price levels; rising expectations about rents tends to result in rising property values, an increasing availability of finance and rising land values. If rent expectations are not satisfied, profits will tend to fall and losses may be incurred, and property values will tend to fall. If property values fall to any extent, land values will also tend to fall. Moreover, where finance has been borrowed on the strength of expected rises in property values, rent income may not cover interest and other costs, and lenders may foreclose. The resulting forced sale of properties will tend to bring values down even further, depressing the market and causing developers and property investors further losses.

Sufficient has been said to make it clear that the extension of contractors forward vertically into speculative development and property ownership is very different to extending vertically backwards into the manufacture of materials and building components, the supply of plant and materials and building components, the supply of plants and similar development. Backward extension may result in stabilising supplies and prices, and increasing the stability of the firm. Forward extension into property development and ownership can entail a new range of risks of an extensive nature but also the possibility of greater and more stable profits.

DEPARTMENTAL EFFICIENCY AND CO-ORDINATION

The problems of departmental efficiency and departmental organisation are not really separable. Each department exists to provide the required amount of services to the market or to the other departments. All the departments tend to be interrelated; a high level of departmental efficiency is only of limited value if it does not match the needs of the other departments. For example, a plant department, however efficient and however low its charges, is of limited value if it cannot supply or borrow plant when it is required. Again, a surveying department is not satisfactory, however efficient, if its staff cannot handle the load of work at other than slack periods, for the cost of credit to finance unsettled accounts may exceed the extra cost of the staff needed to handle the extra work.

Building problems must often be settled rapidly and on the location of the work. The remote headquarters manager tends to be at a disadvantage both from the point of view of information and of the speed of decision. The managements of the various departments must often be entrusted with wide powers. The need to refer back to a high executive level can lead to delay and to heavy costs and to the saturation of an element in the chain of command. On the other hand, Co-ordination is more difficult where the power to take decisions is widespread. In such circumstances it is difficult to prevent waste and overlapping, and to maintain the activities at a full and even level.

The success with which a firm is run is, of course, ultimately measured by the profits obtained. The return on capital is still applicable to a non-profit organisation. But profits are normally computed in a very crude way and do not always provide a very useful indicator of the efficiency of the organisation. The uncertainty of estimating for tendering and the consequential incidence of contract losses and gains results in profits being a particularly uncertain measure of efficiency in the building industry. The solution is to make frequent cost checks, not merely for the contracts, but also for each of the services.

COST CONTROL

Much of the necessary machinery for these cost checks, and indeed for the control and co-ordination of the separate units of the firm, can be exercised through the accounting system if it is properly designed to provide a system of accounts which, as well as providing the normal book-keeping facilities and financial returns, also provides a measure of the economic efficiency of the separate units and a system of built-in guides to economic choice. In other words, what is required is a system of accounts which provides at one and the same time a comparsion of costs at market prices and which creates the conditions of a free market within the firm.(106)

Most of the systems of accounts in the industry are adequate for the allocation of direct costs but fail to provide for the correct allocation of overheads. This is rather a serious limitation because the overheads usually form an important part of those costs which are under the direct control of the firm. It is difficult to measure the efficiency of the various divisions of a firm unless the overheads are correctly proportioned to the various functions carried out. Consequently, management lacks a basis on which to decide the best form of organisation, or the best way to dispose of their assets so as to obtain the greatest returns.

Goods and services will be sold to the public at market prices; internal services are, in fact, in competition with the market, and the justification for providing such services within the firm is an adequate profit margin. These requirements can be reconciled by charging market prices where applicable on internal transactions and by adding a customary profit margin on to the costs for capital items and other services not obtainable from the market.

The actual system of accounts most appropriate will vary with the individual firm and with the relative importance of the various functions it carries out. The basic requirement is that the system should provide indicators of the profitability of any particular productive process or service, and provide a basis for the comparison of the costs of alternative processes and for cost estimating. Most organisations produce some goods and services of a kind which could be obtained from the market and so indicators are essential to determine whether direct production is worthwhile. Comparative costing is necessary to provide a basis for selecting between alternative methods.

The returns from the capital used in each sector can be readily obtained if the capital is charged to each sector of the firm at a fixed rate of interest, and the net interest paid by the sector can be used as a measure of the amount of capital employed in it.[106] Thus the return on the capital used in each sector can be measured by dividing the surplus by the amount of net interest paid and multiplying by the rate of interest. For example, with a rate of interest of 5 per cent, net interest of £500 and a surplus of £1000, the sector would have obtained a net return of 10 per cent, or, including the interest itself, a gross return of 15 per cent. Thus it is possible to check whether each of the various activities of the firm is making a proper contribution to the earnings of the firm. If the return is below the normal expected return the assumption would be that the service was operated inefficiently, and either it should be reorganised or it should be closed, and the capital used elsewhere in the firm where it could be used to earn a proper return. The interpretation of the returns from the contracts is not as simple since they depend on the validity of the tender price as well as on the success of the management, but they still provide a measure of the profitability and provide a starting point for determining the cause of the failure.

The central services are more difficult to analyse since usually the services have no market value, and there is no alternative of purchasing them from the market. But the cost of the services can still be compared with some measure of the service provided, for example, the value of tenders entered, or the number of operatives bonused. A measure of efficiency is obtained by comparing one year with another, or by comparing the results of one branch with another, or one firm with another.

The return on the capital is, of course, only an average figure and a first indicator of profitability. The maximum return on the firm's capital is obtained when the marginal returns from each sector are equal, that is when the returns on each final unit of capital used for each sector are equal. Thus if maximum returns are to be obtained, it is necessary, not only to ensure efficiency in each sector but to move capital from one sector to another until the final increment in each case is producing the same return. This position can often be found by trial and error.

CHAPTER FIFTEEN

THE ORGANISATION OF PROFESSIONAL SERVICES

THE DESIGN TEAM

Attention has already been drawn to the tripartite nature of the construction industry, the division between designers, contractors and producers. The designers, the professional element in the trinity, are themselves divided into a number of separate groups of professions, each dealing with their own segment of the design process. The main division lies between architects, engineers and quantity surveyors; other professions such as valuation surveyors and town planners are also often closely associated. The pattern varies to some extent from one country to another.

Clearly, unlike the contractors and the producers of materials, the designers have an effect on all the costs of construction and on the running costs of the building. Hence the designers hold the key to the economic use of resources in the construction industry. As mentioned earlier, economic design demands a wide and detailed knowledge of aesthetics, of the way buildings are used, of the functioning of materials and structures, of the problems of site erection, and of the costs of construction and running buildings. This is a formidable requirement and clearly few individual designers can expect to be fully competent in all the aspects of building design. It would seem that design needs to be the work of a team. How the team should be organised is a question about which there are many opinions.

THE DESIGN

The function of the design team is to design a building or works which will enable the functions to be performed in the building to be carried out with the maximum efficiency and in the most economic manner throughout the life of the building, and at the same time to provide a building pleasing both internally and externally, and in harmony with other buildings in the area. A long-term view is necessary of the building, of the functions it is designed to serve, of the way the component parts of the building will function and of the costs that will be incurred. Some balance is needed between the fitness of the building for its original purposes and the need for adaptability to meet future needs.

Perfection is no more possible in the field of design than elsewhere and its pursuit beyond a point may lead to inefficiency. In any activity decreasing returns set in after a time, and theoretically there must be an optimum time worth expanding on any particular design. After a time, the additional cost of further design time will be worth more than the improvements to the design. Thus, perhaps, the concept of value against first and running costs should be extended to include design costs. Since design costs probably only account for about a twentieth of first and running costs it will be worthwhile to spend time on design in return for quite small proportional reductions in the combined costs. The time ideally required to carry out the design may vary considerably from one building to another, depending on its complexity and on the degree of familiarity the designer has with the problems to be solved.

Prior to commencing design work it is advisable to agree a clear brief with the client. Studies suggest that frequently the client is vague as to the exact purposes for which he requires the building, the performance parameters, the way these are likely to change in the future and what he can afford by way of initial and running costs. If the design is to be satisfactory, the designer needs to stimulate the client to work out his requirements in clear terms.

Frequently the client is not the future user of the building. He may be building for letting or sale when he needs to anticipate the requirements of likely future occupiers. In large organisations such as local authorities, hospital authorities and property companies the designer is likely to be taking instructions from officials who have little direct contact with the final users. To some extent this is unavoidable because the final users are unknown and even current representatives of their classes may have little idea as to how the building will be used in future. Often officials, not directly concerned with bearing the final burden of initial and running costs, demand facilities and standards which the ultimate users find unacceptable. Moreover, the opinions of officials often conflict, opinions carrying the most weight tend to dominate unrelated to their merits, resulting in unnecessary and unwanted design features. Handling unco-ordinated and unresolved demands tends to take a long time, much design work tends to be abortive and the end result to be unsatisfactory to final users. Where the client does not directly represent the final occupiers, sample studies of the way similar occupiers use buildings and of their systems of value would appear to provide a better basis for preparing a brief than the opinions of the main client's staff.

DESIGN FEES

The problem of how long to spend on a design lies behind the difficulty of determining professional fees. Ideally, the fees should be large enough to pay for the type of staff to be employed, for the time that is needed to prepare an adequate design, and to cover the overheads of the office. The R.I.B.A.[107] found a wide range of design costs for architects' work varying with the type and size of the job. The variations were quite unrelated to the scale of fees normal for architectural work. The figures indicated that at a standard fee of 6 per cent the design of industrial buildings tended to be the most profitable and houses the least profitable. These results were largely related to the amount of detail and repetition, and were not related to the efficiency of the design. Unfortunately, the amount and standard of office job costing was not at the time of the inquiry adequate to provide the basis of a more logical system of charging.

While any scale of fees which is simple enough to be readily understood by all concerned must inevitably be fairly rough, it should broadly relate fees to the amount of design work desirable to produce the optimum relationship between design cost and total costs. The scale should be related to the time required and the overheads of a normally efficient design team. If fees are fixed too low there will be a tendency for insufficient time to be devoted to the design. Clearly, any scale of fees would need to differentiate between different types and classes of buildings and works according to their complexity.

The present system of charging based on the value of the resulting contract is similar in principal to the cost plus type of contract, in which the reward increases as the cost of the job rises. As a result, designers who succeed in producing economic designs tend to be paid less than designers who are less successful. It might be argued that the code of the professional man is a guarantee that the best service will be provided whatever the effect on the rewards. This is a noble sentiment but does nothing to remove the objection to a fee system which penalises conscientious work. Another disadvantage of the usual fee system is that it focuses attention on first costs rather than on total costs. Perhaps a better solution, although far from ideal, would be to relate fees to some measure of physical size, such as floor area, with different scales for different classes of building. There does not appear to be any practical system of payment, corresponding to the value cost contract, under which rewards could be related to the value for money that the design offered. Clearly, any system should be based on a study of the actual times spent on various types of design work. There might be some advantage in the scales including an allowance for specialist design work such as engineering and for the work of the quantity surveyor. This would, in turn, raise problems of the independence of the separate specialist.

A soundly based scale of charges could act as a starting point for rationalising the time spent in the design office, for programming and budgeting the design process, and for determining the need for changes in the organisation and technique. The rationalisation of the whole design and supervision process must be preceded by a thorough study of all that is involved.

The amount of design work clearly varies within building types, even when allowance is made for size. Scale fees should not be applied rigidly but they do provide some guide to clients as to how much design cost to include in their budget. It has been asserted, for example, in Great Britain, that mandatory scales operate against the public interest and that there should be open competition for design fees as in Germany and U.S.A. This point appears to be of less importance than obtaining a design which meets the final client's need at the minimum initial and running costs (Chapter Seventeen). Design fees are a small part of these total costs-in-use and additional fees paid for more design work might often be well worthwhile. The client's problem is in determining which designers will produce the most efficient design and how much design work is necessary to obtain it.

THE ARCHITECT

Usually the architect is the leader of the design team. It is he who receives the commisison to design and supervise the erection of the building. The volume of specialised knowledge necessary for the design of a modern building is so great that the architect usually needs the assistance of specialists to handle some of the design problems. For example, civil and structural engineers are often called in to design the frame of the building, electrical and other mechanical engineers to design the services, and quantity surveyors to advise on costing aspects and to prepare bills of quantities where these are required. Some firms of designers tend to employ or to include partners with the required specialist skills, but problems arise in providing continuity of work for the specialist similar to those arising on the contracting side. In a few countries there are designers who specialise in site construction problems. Clearly, a building design is a compromise between the best solutions for the many functions it needs to serve. If the final solution is to provide the best results the various problems must be considered together since the solution for the various functions tend to react one upon another. The arrangement of space within the building, shape, the form of construction, heating, lighting, and other services, and the first and running costs are all closely related. It is more and more difficult, as available knowledge and the range of materials and forms of construction increase, for an architect to know not only the forms of buildings and their layout and equipment but also the way in which the frame and the services, the environment and the costs all react together, especially if he does not specialise in one type of building. All the many factors to be considered in the design need to be considered at the same time. The most economic results will not usually be obtained if the form and layout of the building is determined before it is known how the resulting problems can be handled structurally and how they can be serviced. It may, therefore, be necessary for the architect to become much more of an executive with a general knowledge of all the disciplines involved, or to abdicate the function of organiser of building design and to become the specialist in form, layout and decoration.

The architect requires a knowledge of the principals and experience in the application of the various factors of design, and information on the materials and components available for use in building construction. In the first field, the specialists assist. The provision of adequate information is partly as question of documentation and partly a question of testing and bringing together experience of performance. Much of the information on technical aspects, size, weight and operation is available from the literature of the manufacturers, and the problem is largely one of bringing the data together in a uniform way so that reference is simplified. Various organisations already go some way to meet this need. Less information is usually available on the way one component relates to another, so that unexpected problems arise in construction. Information on costs is often not as easily available as is desirable.

Perhaps the most serious deficiency lies in the available knowledge of the way materials and components, and forms of construction behave in practice. Clearly, it is not very efficient for each designer to collect his own information on the behaviour of materials and components and on buildings. Not only is his experience likely to be limited but usually the designer loses contact with his buildings shortly after they are completed. It would be more efficient for some central body to take over the function of testing and reporting on materials, components, and types of construction. To some extent services going some way towards meeting these needs do exist in some countries. For example, the standards institutes, examples of which are found in most Commonwealth countries, lay down standards and codes to which materials, components and structures should conform. These are usually based on the experience of producers, users and research stations. They provide a simple coded system for specifying but often are insufficiently related to scientifically based testing and are too general to provide the sort of guide architects and others require. The *Agrément* system discussed earlier seems to offer a possible solution. Some large-scale building users have set up their own laboratories and inspection service, for example, large local authorities and government departments. These test individual materials and components, collect information on building behaviour, and issue bulletins to their own designers on performance. Something of this kind is necessary on a national or even international scale with individual reporting, not just on each type of product, but on each make of product. Only then will architects and other designers be in a position to make a rational choice in specifying the materials, components and forms of construction.

A centralised service is also required to provide information on the relationship between different designs and their construction and running costs, so as to provide architects with some guide as to the best type of solution to their design problems.

Architects tend to produce a special design for every building for which they are engaged. In part, this arises because of the variety of design requirements which most design offices are called upon to handle. But designers tend to produce a fresh solution for each building, even where there is no change in the requirements. It would appear to be far more economic, both of design time and construction time, to treat each design as a step towards the optimum design. Each subsequent design would then follow the same lines but with modifications aimed at improving the less satisfactory features. Finally, the design would be standardised and used when that particular type of building was required. This procedure is already followed to some extent; for example, in Russia large numbers of housing units are erected to the same design. In some countries standard designs can be purchased. There is a considerable degree of uniformity of design on housing estates and often on industrial estates. Systems of building, also, are based usually on a small range of designs. Many types of buildings, for example, schools, hospitals and police stations, are required to provide similar facilities. Often these buildings are separated by considerable distances and it is questionable whether they need to be designed completely individually. Again, while uniformity on estates can result in monotony, it can equally well produce very pleasant results; some of the most attractive parts of cities are composed of buildings of uniform appearance.

The use of system building could have radical consequences for the architectural profession. The use of industrialised systems could further narrow the field of buildings designed by architects for individual building owners. The architectural profession is already by-passed by the building owner when he purchases speculatively erected buildings, when he purchases prefabricated constructions direct from the manufacturer, and when he goes to a package deal firm. In some cases the advice of the architect is sought in designing the basic building but the architect has no direct contact with the client. These methods should lead to the design economy possible from the repeated use of the same design. Perhaps better results could be obtained if the architects produced standard designs which they were prepared to apply to the client's requirements: the services of arranging for the contract and supervising it could be provided in the usual way. Clearly, a special and considerably lower scale of fees would be appropriate in such cases. But one-off designs are expensive to construct as well as to design and, in future, greater uniformity may be expected. The forms of construction used need not be limited to traditional ones but could be based on any system that provided good value for money. Indeed, the CLASP and some other systems of construction have been developed by the clients' architects rather than by contractors. Some systems have been developed by consulting engineers.

THE ENGINEER

There is a large range of different types of engineers in the construction industry. These vary from civil and structural engineers to electrical, heating, ventilating, and other service engineers. Engineers either practise professionally as consultants, or are employed by contractors specialising in the various branches of engineering.

Civil engineering works are usually designed by consultant firms of civil engineers who act as principals in the design and supervision of the work in the same way as architects do for buildings. Often the civil engineering consultants take a larger part in the control and organisation of the construction work than architects. Where architects are needed in the design of civil engineering work their place in the design team is usually secondary. Civil engineering consultants do not usually employ quantity surveyors but prepare the contract documents themselves and carry out their own costing.

Architects usually engage either a consulting structural engineer, or ask a specialist contractor to design the frames of their buildings. Where all that is required is a standard frame of steel or concrete, the architects carrying out the design work frequently merely name the firm to be given the contract and the quantity surveyor sets aside a provisional sum to cover the cost. While this method saves the consultant's fee, the design work has still to be carried out by the contractor undertaking the work and the contract price must include an element for the design work. There is some controversy as to whether consultants or contractors produce the most economic designs. On the one hand, it is argued that the professional consultant provides the most efficient design, that he is free from bias as to the materials to be used, and that he has no vested interest in producing a design to sell any particular manufacturer's materials. On the other hand, while it is agreed that the contractor usually works in only one material, it is said that his experience in erecting as well as designing frames result in his producing very economical designs, and since he is in competition in the long run, if not for a particular contract, he has every incentive to produce a competitive job.

On the services side it is perhaps less usual to go to a consultant although, again, consultants are often employed on large complicated jobs. Again, there is little agreement as to whether it is better to employ a professional consultant or to put the work out on the basis of design and construct to a contractor specialising in the field.

As already mentioned earlier, it is in their relation to the design and contracting of the specialised work, mainly engineering, that the practice of architects differs most from one country to another. Under British practice the architect appoints the consultants and usually nominates the engineering sub-contractors. The American architect, on the other hand, generally leaves the main contractor to appoint the engineering, as other sub-contractors and these compete for the work. Each sub-contractor works out his own design within the architect's general requirements. While the architect has the right to approve the design, it is not thought that in practice he has much control over the detail.[94]

THE QUANTITY SURVEYOR

The quantity sureyor has developed as a building accountant. His function is to measure the quantities of materials and labours from the plans and specifications prepared by the architect, provide estimates of the likely costs, and, later, to measure any differences between the job as tendered and as built, and to evaluate such differences on the basis of the unit rates written into the priced bill by the successful contractor. As a professional man it is his function to act as the honest broker between the client and the contractor. Under British practice a quantity surveyor is appointed for all substantial jobs.

The limitations of the traditional methods of taking off quantities have already been considered and it is clear that there are alternative methods which might provide a basis for an improved service from quantity surveyors. Some form of operational bill might provide a better base from which quantity surveyors could provide cost advice as well as a better basis for contractors to use for tendering. It seems logical that if quantity surveyors are to take off the building work, they should also take off the engineering work as they do in Scotland and some other places.

Quantity surveyors are now paying more attention to costing as distinct from accounting.[(109)] In the past the costing aspects of their work were limited to little more than trying to predict what the tender price might be. More recently they have been concerned with the development of cost-planning techniques. These techniques include both methods to aid the designer to distribute the expenditure allowed for the building reasonably over the various components of the building, and methods aimed at finding which alternative form of construction provides the best value for money. These developments suggest that the quantity surveyor will in the future need to be closely integrated in the design team, instead of, as often in the past, being a separate consultant who provides a service when the design is complete.

If the quantity surveyor is to evolve in this way he will need a broader training in statistics and economics to fit himself for the role of building economist. It would appear, however, that both cost planning and design-cost appraisal are only practicable on the basis of the continuance of some sort of bill of quantities, together with the system under which the contractor prices each element of work separately. Without a reasonably detailed basis of prices there would be no basis on which to estimate the likely prices at which contractors would be prepared to contract for the work. Little cost planning or cost appraisal could be carried out if contacts only quoted a single contract figure. In such a situation the only source of price data would be the contractors' own records and it seems doubtful if they would allow access to these, even for the purpose of preparing a statistical summary of price data. In the absence of detailed price data it is difficult to see how designers are to know whether they are designing with due economy. This is the situation under the American system. There the lack of price information inhibits the development of design cost techniques. It may be that this absence of price data and the resulting absence of price control offsets the apparent advantages of the greater element of competition for contracts. It has been suggested that the weakness in cost control may be part of the reason why the architectural profession is losing ground in America.[108]

Even in countries which do not follow the British system there is often some system of quantity surveying service. For example, while in America there are no bills of quantities there are quantity surveyors who provide a checking service to contractors.[110] Again, in France, tendering is on the basis of plans and specifications but there is often a bill of rates to provide a basis for pricing variations.[111] In one or two districts bills are prepared by a quantity surveyor appointed by the architect but they do not form part of the contract.[111] Quantity surveying firms exist either to prepare quantities for contractors, or for architects. Some firms will work for either, but they are not independent professional consultants and only represent the interests of one party to the contract.[111] Generally, there is no quantity surveying profession, as it is known in Britain, outside the Commonwealth and other formerly British territories.[93]

The spread of proprietary systems of building, package deals and other forms of negotiated contracts reduces the number of occasions upon which the test of competition can be applied and the volume of detailed price data which is available. The client is particularly in need of the type of cost service provided by quantity surveyors in the negotiation of contracts, for the only test of the reasonableness of the price is the corresponding price paid under competitive tenders. The use of negotiated forms of contract should not, therefore, be pushed so far that the number of competitive tenders is inadequate to provide a basis of comparison. Of course, considerably better use could be made of the information from priced bills if these were analysed centrally; the modern computor provides the means for a rapid and thorough analysis of bill data and indeed for the preparation of bills.

COMPREHENSIVE DESIGN TEAMS

If value for money is to be provided a multi-discipline team is necessary. The design needs to be the joint product of architects, structural and service engineers, specialists in erection and site organisation, quantity surveyors, and other specialists in predicting cost relationships. Ideally, a team containing all these disciplines would be the best solution. Consultation between them would be closest if they were united in a single firm. But different types of buildings tend to make different demands on the time of different specialists and it might be difficult to keep all the specialists fully occupied if they were in a single firm. Moreover, a single design team might lack the stimulus to the development of new ideas which might be provided by participation in many different teams, the specialists in which would tend to use different solutions and thus throw up different problems for solution by other members of the team.

In France a large design organisation, *Omnium Technique d'Habitation,* has been set up to provide consultancy services in engineering, other technical aspects of design, planning and economics. It will undertake design work, including the develoment of new systems of construction, contract negotiation and site organisation.[93] In Britain a government-sponsored National Building Agency has been set up to provide somewhat similar services.

PROFESSIONAL SERVICE TO SMALL CLIENTS

Most professional services have in the past catered mainly for the needs of clients with a substantial building problem, generally in relation to a new building. Despite the scale of small extensions, repair and rehabilitation work, professional practises have not generally been organised to provide the advice and services which consumers need for such work. The client with a small building problem such as the installation of heating appliances, structural repairs and alterations, and small additions usually obtains the advice required from a contractor or supplier. However reputable, such advisers are concerned to sell their goods and services and can hardly be expected to seek out the most economic solution and contractor, even if they had the knowledge, especially where the business would then go elsewhere. In fact the demand for such work, especially for double-glazing, the supply and installation of mechanical appliances and extra rooms, has given rise to the mushrooming of firms who specialise in such contracts. Many undertake intensive advertising and selling and some are reported to provide poor value for money. If clients are to obtain unbiased advice and competitive contracts they need professional advisers, although a service more informal than that provided for new building works on a large scale. Some professional firms have seen this need and have set up High Street building consultants providing a flexible service. They will provide, according to need, building advice on the best solution, choice of appliance, materials and contractor, provision of design, specification and cost estimates, supervision, inspection and arbitration. The development of such services could provide a large amount of employment for professionals in the building industry.

PART IV

Planning and Design

CHAPTER SIXTEEN

PLANNING AND STRUCTURAL ECONOMY

DESIGN ECONOMY

The object of the design team is to provide good value for money. This does not mean cheapness but economy. Saving cost by cutting standards does not necessarily improve the economy of a building, although it may cheapen it. The most economic building is the one which provides the values required at the lowest cost. Values and costs need to be considered together. Value is derived from appearance and function; together they are judged against cost. This trinity is implicit in all evaluation of buildings and works but it is a difficult concept to apply.

Design and development economy are both intimately bound up with standards of amenity. It is simple to reduce costs if the standards of construction and finish are lowered, or if the amount of space is reduced in relation to need, but this is not design economy. This involves a reduction in costs without a reduction in amenity. Design economy is only likely to be achieved if costs as well as amenity are considered jointly throughout the design process. In relating design to costs a long-term view is taken, amenity and cost being considered over the life of the buildings.

The need for cost planning implies the availability of alternative methods of construction and materials. It is because of the wide choice of materials and forms of construction which are now available that the need for economic evaluation of building design is so necessary.

PLANNING ECONOMY

Clearly, the first step is a thorough appraisal of the way the building is to be used, not just initially but as far as can be seen over its life. A study of the functions to be performed in the building will indicate the relative advantages of different shapes and floor areas, of arrangements of space, and of the services and components which are needed. Such a study may, in some cases, indicate better ways of performing the functions. Where the functions to be performed are very complex, long studies by people of many disciplines are often necessary. The final plans must inevitably be a compromise, since functions change, and close tailoring to the needs for current functions may reduce adaptability and reduce the overall long-term efficiency of the building.

Modern materials such as steel and concrete are flexible in their use in relation to building needs and impose very little limitation on planning. But, of course, the further the materials are pushed to their limits the greater tend to be the cost of using them and, for example, after a point the cost of increasing spans rise very steeply.

Environmental needs also impose limitations on planning freedom. For example, natural daylighting cannot be provided at an adequate intensity at points deep in a building. If natural lighting is to be provided, room heights must be increased as depths of room increase. Extra glass increases heat losses in winter and gains in summer as well as admitting more light. Natural ventilation admits noise as well as air.

The shape of a building affects both the structure and the environment, and hence both the first and the running costs. The minimum area of enclosing surface to volume is given by the sphere but this is not a very practical shape for a building. Circular buildings are used for storage and other regular-sided buildings have been used. More usually, buildings are rectangles for which form the square gives the minimum enclosure to volume ratio. The greater the length-breadth ratio the more materials are required to enclose the space, and the greater the surface area through which heat losses can take place.

Of course, the various forms of enclosure have different costs and it might be more economic to reduce the ratio of roof to wall area, even if the total area of enclosing surface was increased. Again the cost of walling may vary with its form. For example, in the case of terrace housing external walling will tend to have a different cost from party walls. The most economic width to depth will be where the ratio of each is in inverse proprotion to its costs—costs including those for running as well as initial costs.

Both running costs and first costs tend to increase with increases in the length—breadth ratio.[(112)] But other factors must also be considered, for example, lighting, noise and circulation within the building, and the effect of these may be to swing the advantages in favour of some other form or shape. For example, while staircases and lifts add to the costs both of erection and running a building, provision of these may be less expensive than the cost of providing and running horizontal communication links.

Cost is nearly always a factor of importance in the final planning of a building. In the past it has intruded mainly as a ceiling on the limit of expenditure. Today, with such a large variety of possibilities for each aspect of design, it is a factor to be considered as each detail of the plan is determined. It is no longer accepted as adequate that costs can be cut by reducing the volume or area, or by reducing the standards of fittings and finishings. Cost-planning techniques are now available to guide the designer in achieving a proper balance of expenditure on each part of the building and to assist him in obtaining good value for money.

While the first item to be considered in the planning of a building is the size, generally savings in costs are far less than proporitonal to reduction in size.[112] Changes in area may result in only small changes in the wall and roof areas and the amount of frame, and will tend to have little or no effect on many of the items of construction, for example, the heating plant and the fittings. It has been shown in the case of low-cost housing in Britain that, on average, the percentage change in cost is about half the percentage change in area.[113] While it is uneconomic to waste area, it does not usually pay to try to reduce the floor area to the minimum that will suffice. Space demands tend to increase both as society becomes more sophisticated and as families and institutions get older. Moreover, adaptation and conversion to new needs tends to be more difficult where the space is very limited. In the long run it may be economic to be fairly generous with space. While ceiling heights are generally less limiting, future needs have to be considered in determining their height.

Space savings are still important even though the reductions in costs are less than proportional. Up to a point, it is probably easier to make savings by reducing the floor area than in other ways. Moreover, often the functions to be performed in the building can be carried out more efficiently where the space is not excessive. Where intercommunication is important it is often an advantage to arrange the working spaces close together. In this way the process can be performed more efficiently and at the same time the floor area reduced. Often far too much space is taken up by the commuication links. For example, a study of some blocks of flats showed that the ratio of access area to gross floor area on the upper floors varied from 9 to 29 per cent; studies indicated that 15 per cent should have been quite adequate.[114] Again, studies of the use and planning of floor space for schools indicated that on average floor areas could be reduced by 40 per cent without reducing the teaching area. As a result of this and similar economies, the real cost per school place was reduced by 20 per cent without sacrificing the real standards of accommodation.[115]

Costs are affected not only by the amount of space but by the form of enclosure. Usually the plain rectangle is more efficient than any other form of enclosure. It is not sufficient to minimise the area of enclosure in relation to the floor area; simplicity in construction is also important. Cornering usually adds to the costs and right angles are usually cheaper than other forms of angle. While a regular, many-sided shape may have a smaller wall area than a square, the saving in wall area costs may be more than offset by the additional costs of forming a larger number of more difficult angles. The relative position will depend on the form of construction. Moreover, the space in rectangular buildings may be more useful than in other shapes, especially if the buildings are small. Projections usually add disproportionately to the costs in relation to the floor area which could be added by increasing the size of a rectangle. But, of course, wall enclosure is only one of the factors to consider, since additions to the wall area may be worthwhile in order to obtain a saving in the costs of the frame, or perhaps in order to be able to make a fuller use of natural lighting. The effective floor area may be rather less than the actual floor area. Some effective floor space is usually lost around the columns, the amount depending on the way the building is used. For example, in a factory with very large machines, or where large units are assembled, the number of columns and their spacing may have a considerable effect on the amount of production which can be carried on, and the cost of the effective floor area per square foot may be lower if wider column spaces or even no columns are used, even if a more expensive frame becomes necessary. Again, on an expensive city site it may pay to use an expensive form of walling and perhaps expensive partitioning, if, as a result, their thickness can be reduced and a gain in floor space obtained.

Shape also affects the costs of the components of the building. Tall, narrow window openings are often cheaper to construct than wide, low openings of the same area. Roof shapes vary considerably in cost, depending on their form of construction. In Britain, hipped roofs were found to be over 10 per cent cheaper than gable roofs.(116) Pitch is also of importance; it can make as much as 33 per cent difference to the costs.(117) Dormer and mansard roofs are usually more expensive than the normal pitched roofs. But, of course, it may sometimes pay to use an expensive solution for one component of the building in order to obtain savings elsewhere. Roofs can have a wide range of costs as studies of, for example, factory roofs(112, 118) and the roofs for flatted blocks(119) have shown.

Savings can often be achieved by arranging for a dimensional pattern convenient for the material to be used. For example, savings can be obtained by centring floor joists to fit in with the optimum spacings for the floor boards and with the widths of the plasterboard to be used on the ceiling below.(116) This type of planning is, of course, related to the use of modules, which has been considered earlier. The relationship of services can also be of importance. For example, it was found that in two-storey housing it was cheaper to place all the sanitary appliances on the ground floor than the usual arrangement of placing the bathroom over the kitchen.(120)

The number of storeys in a building usually has an important effect on costs. Usually the greater the number of storeys the greater the costs per square foot. The relationship of costs for one- and two-storey buildings tends to be largely dependent on the amount and costs of access to the first floor. For some forms of housing there is little difference between the costs of one- and two-storey buildings, the balance one way or another depending on the relative costs of staircases, walls, foundations and roofs. In the case of multi-storey housing, costs in Britain tend to rise rapidly at the point where load-bearing structure gives way to framed structure and lifts need to be introduced. After this point the rate at which costs rise with the number of storeys tend to fall.(121) But it is difficult to generalise; the relative costs for different numbers of storeys varies considerably with the forms of construction and with the relative prices of different types of labour and materials. As a result the price relatives vary considerably from one country to another. The increase in prices with the number of storeys for flatted blocks is far greater in Britain than it is in America,(16) or on the Continent. Part of these differences is a result of differences in the prices of local supplies of labour and materials, and of differences in the types of buildings representative at different storey heights. The effect of storey height on costs depends on the type of building being considered. The evidence for Britain is that costs rise with the number of storeys far more for flatted blocks than for factory building.(112, 121)

Planning studies of high blocks of flats in Britain have revealed considerable differences in the efficiency of different designs. For example, it was found for a sample of blocks that the ratios of gross floor area to external wall area varied from 0·9 to 1·6 for rectangular slab blocks and from 0·7 to 1·4 for other types of blocks. Lift installation costs were found to vary by a factor of 1 to 4 per flat.(114) This variation was largely the result of the number of flats per floor served by each lift. Speed, the degree of automatic control and the number of storeys all affect the costs of installing and operating lifts.(122) Sometimes quite unnecessary features are introduced into design. For example, a second staircase in a block of flats rarely fulfils its prupose of improving escape facilities and has been found to add significantly to the cost of a dwelling.(123) Some shapes of block appear to be far more efficient than others. For example, double tower blocks can be designed to give the highest number of flats per floor to each lift, low ratios of common access area, and high ratios of floor space to external wall area.(124)

Of course, different plan types have different aesthetic and social values. While such values are mainly a question of personal judgement and cannot often be costed, they are nevertheless of importance and must be taken into account. So while one solution may be more expensive than another, the extra expense may be worthwhile in view of the better aesthetic or social values associated with it. For example, while staircase access to blocks of flats is, in Britain, 10 to 15 per cent more expensive than balcony access,(125) the extra costs may be worthwhile in order to secure the greater convenience which staircase access is thought to provide.

STRUCTURAL ECONOMY

Structural economy depends on the use and choice of materials, and with methods of erecting them. Thus it is concerned with the costs of labour and materials, plant and organisation. Again, while in the first place it is convenient to examine the economics of each material separately, in practice, constructions need to be examined as a whole so that the mutual reprecussions of the various design decisions can be considered.

The choice of materials depends both on the prices and on the properties. Relative prices change and with them the solution which is most economic. Some discussion on the interplay of labour and material costs was given earlier in considering traditional and non-traditional forms of construction. Illustrations were given of the way materials used for construction have changed over time with changes in the supply and cost of materials, labour and transport. But it is not sufficient merely to choose the cheapest material, it is also necessary to use its characteristics to their fullest advantage. It is wasteful to use materials with qualities over and above what is required in the given situation. For example, a study of the design of some multi-storey flats in Britain indicated that concrete was often used below its full capacity and that in some cases three times as much concrete was used as was necessary.[114]

The way in which material is put together is equally important as the exploitation of its consequences. Savings can be achieved by the use of designs which allow for the repetition of simple operations, and for the concentrated use of mechanical and other plant. The cost of providing and fixing shuttering for concrete is often high; sometimes it costs more than the concrete it encloses. The simpler the design of the shuttering and the more often it can be used, the cheaper will be the finished price of the concrete. Again, some bonds for brickwork are much more expensive to lay than others.

As mentioned earlier, the preparation for an operation and clearing up after its completion tend to take much the same time whether the operation itself is long or short. It is, therefore, advantageous to reduce the number of operations necessary for a given job as much as possible, so that a labour gang can complete their work in as few operations as is possible. Again, it is an advantage to complete similar operations over as short a time period as possible so that the plant can be used intensively over a short period.

For low buildings with narrow spans, such as residential building, small shops and offices, schools and other buildings on a small scale, some form of masonry construction or brick usually seems to be the most economic. In England, the cheap fletton brick tends to result in such forms of construction being very economic. The use of bricks for these purposes is challenged by concrete blocks in areas at a distance from the brickworks, when transport costs tend to result in the brick being less favourable. Light-weight concrete blocks are sometimes worth using in order to obtain a wall with a better thermal insulation value. The concrete blocks are in themselves cheaper per unit area to lay than the smaller brick, but are being challenged in this field by other types of clay products, for example, hollow clay blocks and double bricks of conventional length and thickness but joined in breadth to form the inner and outer skin at one laying. Considerable use is made of large clay units in some European countries. The deciding factor in determining the most economic form of walling is often the appearance. Any saving from using a substitute material may be more than offset if the appearance is unacceptable and the wall is rendered. In some areas, however, climatic or other conditions have resulted in a general practice of rendering walls and in such cases the use of cheaper forms of walling unit is usually economic. Concrete blocks have shown far more consistently favourable results than brick for the inner skins of walls where a covering of plaster or some other material is usual.[120] In countries where timber is cheap, for example, America, it is the most economic solution both for framing and cladding the walls. In less developed countries, where timber and suitable clay for brick-making are not available and appearance is not too important, locally-made concrete blocks often offer the most economic solution. Where labour is plentiful and cheap and the climate is reasonably dry the older forms of rammed earth walling may still provide an economic solution,

For larger buildings some type of framework is usually necessary and the choice usually falls between steel and concrete, although for some types of large buildings, where the spans required are small, for example, blocks of flats, a cellular structure of load-bearing walls provides the most economic solution. While concrete frames have tended to be favoured in Europe as the more economic, steel is more general in America and in some other countries. Again, the most economic solution depends on the current local levels of prices for materials and labour. Recent advances in the design of structural steelwork may lead to the use of more steel in Europe. The choice of claddings for a framed building is, of course, very wide since they need not also carry any loads. Often, however, brickwork is still an economic choice although its load-bearing characteristics are not required. Light-weight claddings, such as curtain walling, often prove very expensive, especially if account is taken of the future costs for maintenance and cleaning.

Generally, prefabricated structures and wall claddings have not proved themselves particularly economic and their use has been confined to buildings for which cost is not of first importance. But it has been pointed out that often the prefabricated components are not mass-produced but a one-off order for a particular building. The use of prefabricated units will become worthwhile if designers are prepared to incorporate units chosen from a limited range for which the demand is sufficiently regular, and on a sufficient scale for manufacturers to produce for stock. The acceptance of standard dimensions for room heights, window and door openings and for planning modules would assist this process. While the acceptance of standards of this kind must place some contraints on planning freedom, they need not impose any serious practical limitations on design.

So far there appears to be no evidence of the overall superiority of any particular system of building, and it may be that the site-produced systems will prove as successful in the long run as the more elaborate factory systems. What is lost from the greater productivity to be expected from production in the factory may be made up by the saving in the costs of special equipment and in the costs of transport. Systems of construction, such as the battery system,[49] which require little or no special plant equipment, can clearly be used equally satisfactorily for a wide range of building designs and might well add to the range of design methods which architects might use in preparing their designs.

Where systems require an elaborate kit of parts of special design, they are not likely to be suitable for the designer concerned with the design of one or a small group of buildings. They may, however, provide a useful solution to the designer who is able to use either a standard building, or a building composed of standard parts on a large scale. This line of approach has been followed in the case of school and barrack buildings; clearly, it has a use in the housing field.

While the new developments in prefabrication, system building and other forms of industrialisation add to the opportunities open to the designer, they do not really change any of the basic principles of design. The need to exploit fully the characteristics of the materials, to study the impact of one aspect of the design on the others, and in particular on the cost of labour and materials, remains. Practically no design rule can be applied universally; its application depends on the characteristics of the particular design problem. For example, while it can be wasteful to incorporate too much cement in concrete if as a result its strength is greater than is required, the extra costs of handling lean concrete can be greater than the saving from reducing the cement content. Again, the combination of the most economic of structures for inner and outer leafs of cavity walls may not produce the most economic cavity wall. Savings in construction may produce a finished unit of construction with a quite unacceptable appearance or amenity. Long-term cost may outweigh savings in first costs. The costs of different types of labour and materials varies from one place to another and from one time to another, and consequently the most economic solutions will vary from place to place and over time.

DESIGN AND CONSTRUCTION

As some of the examples have shown, design economy is not just a question of minimising the quantity of materials or of choosing those with the minimum costs, but of producing a design with the minimum total erection and running costs in relation to the amenities provided. A reduction in the amount of materials is not necessarily accompanied by a reduction in the amount of labour. In fact, the amount of labour required may actually be increased. For example, while the shortest route for a sewer may be diagonally under the building, the use of this method may increase rather than reduce the costs of construction. This might occur, for example, because of the resulting interference in the construction of the foundations, which might be otherwise carried out as a continuous operation. Moreover, if the sewer passes directly under the building it must be built to withstand the effects of settlement and might be expensive to replace or repair if the need arose. Again, it might be cheaper to use beams and columns of equal dimensions for each storey, even though the loads to be carried are far less on the upper floors. The extra costs of the unnecessary materials might be less than the costs of changing the dimensions of the formwork. It might pay to use precast concrete where *in situ* concrete would be cheaper, if, as a result otherwise idle machine time was reduced. In another situation more expensive precast concrete might be worthwhile in order to avoid bringing an additional machine to the site. Again, while changes in the texture of a surface may not increase the amount of materials, it may add considerably to the cost, perhaps because of an increase in the number of operations, or perhaps because several additional sub-contractors need to brought into the contract. Even changes in the use of the same material may add considerably to the costs. For example, even slight changes in the profile of a wall of *in situ* concrete will involve changes to the shuttering, which alteration, apart from being expensive in itself, causes a break in the operations.

Changes in design usually have some effect on appearance, if not on the way the building functions, and it may, of course, be worthwhile to accept the additional costs in order to enhance the appearance of the building. Often, however, changes in design detail of the kind under discussion have little effect on appearance, or little effect in relation to the additional costs involved. Frequently, however, the designer has no conception of the effect the detailing of the design is having on the costs of construction. Under the British system of bills of quantities this tends to arise because costs advice given to the designer is often based on the unit rates for different types of work and these reflect the volume or area of the units required and their average price and are not related to the particular difficulty of individual jobs. Thus, any comparison betweeen the prices of *in situ* concrete walls for a multi-store building in which the walls for each storey were identical and the costs if the wall profiles and openings changed from floor to floor would only reflect differencs in total concrete and not the differences in the costs of formwork, labour and plant, and it would be likely to be the costs of these which would be important. In countries in which there was no system of priced bills there would, of course, be no readily available basis on which to estimate the costs of the difference in design. It is argued that if bills were laid out in the form of operations the work would be costed in accordance with the labour and other costs actually involved, and not in accordance with the materials required. Alternatively, some idea of the costs involved in different designs could be obtained by examining the way the work might be carried out by programming and costing out the labour gangs needed for the period required, together with the overheads on a similar basis, and by adding in the cost of the materials. A good deal of understanding of the construction process, and of the organisation of building work is necessary to obtain a reliable comparison on this basis. This is usually beyond the competence of the designer or his quantity surveyor, although, of course, as argued earlier, it might be that greater efficiency could be obtained in building construction if someone in the design team had this ability. In the absence of such knowledge in the design team, such estimates can only be obtained by calling in a contractor to advise on the construction problems of design alternatives and on their comparative costs. This is usually difficult unless the contractor is to be given the job, in which case the contract price can only be negotiated.

The cost of a building operation consists of the costs of labour, of plant and other direct overhead items, of materials and of general servicing costs (overheads). The cost of labour consists of three parts, preparation, the task itself and clearing up. Since preparation and clearing up often take much the same time whatever the length of the operation itself, it therefore pays to arrange the work in as few continuous operations as possible. The labour costs consist of the wages paid to the members of the labour gangs from the commencement of the preparation until the completion of the clearing up. There is usually no saving from small breaks in the needs for labour, since it is difficult to find work for operatives for short periods. Similarly, the cost of the plant must be reckoned in terms of the days during which the operation is being carried out; again, it is difficult to find useful employment for plant during short breaks in an operation. The cost of the plant will be the daily hire charge for the period it is on the site for the operation, together with the costs of transport, setting up and taking down; the latter charges may be lower if the plant is already on the site for some other purpose. In some cases, for example, lifting plant, it may be possible to utilise the same piece of plant for more than one operation. Thus in estimating the costs of an operation it is necessary to consider the way in which the operation fits into the programme of construction. Many of the general overheads are related to the overall site time for the whole building and it may, therefore, pay to carry out operations in a time shorter than optimum for the operation itself in order to reduce the overall duration of work on the site. Only the material requirements can be estimated directly from the volume or area of construction. Even for materials the costs per unit may vary according to how and where they are obtained. It may, for example, pay to purchase ready-mixed concrete, or some other prefabricated material, even at extra cost, in order to reduce the overall time of the operation or in order to free plant or labour for some other operation.

Thus economy in design can ony be achieved if the design team has a full understanding of the effect of the design alternatives on the erection process.

While the operational bill is no substitute for such knowledge, it does help in two ways. Firstly, quantity surveyors should be able to acquire a far better understanding of the erection process through the preparation and analysis of operation bills than is possible when working with conventional bills of quantities, and thus be in a better position to advise the architects and other members of the design team. Secondly, the operational type of bill should provide contractors with a far better picture of the implications of a design for the erection process and provide a better basis on which to estimate the costs of construction. Thus the price quoted should reflect the costs of constructing the building as designed rather than the costs of constructing the generality of building. As a result, designers would have some incentive to study the design construction economies, and the costs of uneconomic designs would be less likely to be reflected in the general level of building prices.

If the full advantages of the operational bill are to be obtained, certain other changes in practice would be necessary. The full design details would need to be available at the time the contract went to tender, and perhaps the contractor should be allowed longer to prepare his tender. There might need to be some limitation on the number of contractors invited to tender so as to reduce the national waste from a large number of contractors programming, costing and submitting unsuccessful tenders. Generally, perhaps there would need to be more time allowed for the preparation of designs and for planning and programming the work. Compensating savings should be achieved in the time needed for construction. With better planning the need to make changes in the design after work had started should be greatly reduced.

CHAPTER SEVENTEEN

COST PLANNING AND COSTS-IN-USE

COST PLANNING

In Britain and those parts of the Commonwealth where quantity surveyors practise, methods of cost planning are being developed. Usually cost planning takes one of two forms, elemental target cost planning and comparative cost planning. The first method is aimed at providing a guide to the best distribution of costs over the various parts of a building, and the second method is more concerned with providing techniques to guide the designer in choosing between alternatives.

Designers usually have a cost figure in mind in preparing their design. Usually a target cost figure is set in consultation with the client. This is determined rather loosely in relation to what the client can afford and the quality of building he requires. Sometimes the target is set as a firm maximum and sometimes merely as a general guide. The cost of a building is usually critical for a developer and is often important for other types of client. The maximum price is often laid down for public buildings and can not be exceeded, or only with penalties. Once the design is complete it is difficult to reduce the cost without either complete redesign or reducing the standards of fittings and finishes. On the other hand, if the design proves to be a great deal cheaper than necessary it is difficult to improve space standards and layout without complete redesign, and much additional expenditure on the finishes and fittings may be unnecessary and provide poor value for money. Hence the advantage of being able to plan the expenditure on each element of the building before design starts. This is the purpose of the elemental target type of cost planning.

However, while elemental cost planning sets targets for the cost of each element of the building, it does not provide much guide as to how to maximise the returns in terms of value for money for each design feature, or for the building as a whole. In order to achieve this a guide is required to the design alternatives available for each design feature. For each alternative some guide is required as to its cost and the consequences of its use. This is the purpose of the comparative cost planning type of technique.

ELEMENTAL TARGET COST PLANNING

The overall cost target is usually determined in relation to the type and size of the building required, often on the basis of cubic capacity or superficial area. The area or volume is costed at a unit rate determined from experience with previous buildings of the required size, type and quality. The figure is then adjusted to relate it to the client's cost requirements, size and standard being adjusted accordingly.

The overall cost target is subsequently divided into a series of separate targets for each element of the building. For this purpose a building element is usually a unit of the building which performs the same funciton whatever the design or construction. Usually there are around thirty functional elements but the number varies according to the nature of the building.(126) There are usually separate elements for the walls, roof, partitions, floor finishes, and so on. Naturally, the division broadly follows the form of construction and would tend to differ as between traditional and some non-tradiitonal forms of construction. No firm division into cost elements can be made until the broad nature of the design is clear.

Usually the designer will have prepared his sketch plan and have provided information on areas and storey heights, on probable methods of construction, and on standards, before any attempt is made to prepare the cost plan. The cost plan is then prepared by taking the cost analysis of a similar building and adjusting it for differences in design and in market conditions.(126) The results of the analysis are then usually expressed as target prices for each building element in terms of the prices per square foot or square metre. The architect can then prepare his working drawings with a knowledge of the price for each element which the cost target for the building allows. As the drawings are prepared they are cost checked on the basis of approximate quantities. If necessary, adjustments can be made to the design and specifications to bring the costs into line with the cost targets, or agreement can be obtained to exceed the target cost. When the tender prices are received they are analysed and compared with the cost targets.

COMPARATIVE COST PLANNING

This method, too, starts with an analysis of the probable level of the cost of the building in relation to the requirements. The sketch plan is not, however, prepared by the designer in isolation but in conjunction with the quantity surveyor who prepares estimates of the effect of different solutions on the costs. When the final sketch plan is completed, the cost estimates are brought together to provide an overall picture of the design as a whole. The information is presented systematically as a comparative cost plan. This sets out the estimated costs for the individual elements of the design together with the estimated costs of the alternative methods which might be used. It thus provides a design guide for use in preparing the detailed design. At this stage further alternatives may be considered and cost analyses will be provided. Again when the tenders are received the tenders are analysed and compared with the estimates in the cost plan.

Comparative cost planning thus comes much nearer to the object of analysing the cost consequences of design decisions. It appears to be capable of development to provide a full design cost service. If this is to be achieved too much weight must not be given to costing by elements, since the consequences of design decisions are far wider than the conventional costing elements. Comparative cost planning can also be combined with target cost planning where cost yardsticks also have to be met.(127) So far comparative cost planning appears to have been applied only to first costs. Running costs are as important as first costs and the two types need joint consideration if the best value for money is to be provided.

THE DURABILITY OF BUILDINGS

As mentioned earlier, buildings are very durable; if they are properly maintained they will often last for centuries. Generally, it is only the non-professional buildings in primitive societies which have a short life. Most buildings have quite a long physical life even if their maintenance is indifferent and, in fact, generally even the cheapest forms of construction have a physical life of 50-60 years. By the physical life must be understood the life of the identity of a building, not the life of every physical part of it. While the main part of the building will endure, many of the component parts will need repair and renewal; some at quite short intervals. The length of time for which it will be worthwhile to continue to repair and renew parts of a building will depend on how well it meets the needs of the functions to be performed within it.

In the economic world bygones are for ever bygones. Past expenditure is of no importance, and only the relationship of future expenditure to future values is relevant. A building is worth repairing and renewing if the vaue for money obtained in this way is greater than that which could be obtained by demolishing the building and erecting a new one. For example, it is clearly better to spend £100,000 in altering and repairing an existing building than £120,000 in building a new one if the converted building is as satisfactory as the new one would be and will provide equal service over the same period. Conversely, it would be better to build a new building at £100,000 than to spend £110,000 to convert an existing building and this would be equally true even if the existing building had only just been completed at a considerable cost. Thus it could be sound economics to demolish a recently completed building. Such a situation could arise where the need for which the building had been erected had ceased, or where as a result of inadequate forward planning a building was erected in the wrong place. In such cases it might be quite impossible to convert the building to a new use at a cost worth paying in relation to the value which would be obtained.

It is not, of course, possible to consider the building by itself; it must be considered in relation to its site and associated works. The gain from rebuilding may come from developing the site in a new way, perhaps for a different purpose. Often, particularly in cities, the value of a new building to a different user exceeds the cost of erecting the new building by more than the value of the existing building. When this happens the price offered for the site exceeds the value of the building in its current use and the entire property is likely to be sold for demolition. As cities get larger there is a tendency for more and more building users to wish to acquire a central position. Since the central positions are already in use the prices offered rise from sheer scarcity and it becomes economically advantageous for those for whom a central position is of limited importance to sell out to those for whom it has more value. As the town expands, the pressure for central sites tend to increase and the rate tends to increase at which buildings are demolished either to make way for users with more urgent needs, or to allow sites to be developed more intensively. Thus, in such situations the economic life of buildings tend to be rather short. However, a change of use does not necessarily involve demolition since conversion may be more economic.

Most building structures have a long physical life if they are erected on sound principles. Even buildings of such materials as timber, cob and pisé materials which are not always very durable, can have a long physical life, in suitable climates, if they are properly constructed and maintained. Usually the costs of maintenance are not so great as to shorten seriously the economic life in terms of the physical life. The economic life is more usually shortened because of the economic posibilities of meeting new needs by rebuilding. Even so, many buildings have a life of 60 years and often upwards of 100 years.

Thus building may become obsolete for three different reasons, because of physical deterioration, technical obsolescence or financial obsolescence.[128]

BUILDING RUNNING COSTS

The cost of construction is only the first cost incurred for a building. Over its life it needs to be maintained and serviced. Maintenance will include repairs and decoration, the renewal of some of the components, and perhaps some conversion to keep the building in line with current needs. Servicing includes heating, ventilating, lighting and cleaning. Over the life of a building as much may be spent on maintenance and servicing as the building originally cost to erect. If the construction cost is spread over the life of the building in the form of a mortgage payment, the annual amount is often about equivalent to the average annual costs of maintenance and servicing, the proportions varying with the type of building and the local levels of costs.[128] For example, estimates for a light factory building gave the equivalent annual costs of construction as about 50 per cent of the total equivalent annual costs, maintenance and fuel for space heating each accounted for about a eighth of the total, and fuel for lighting and service personnel for another eighth[129] (Fig. 17.1).

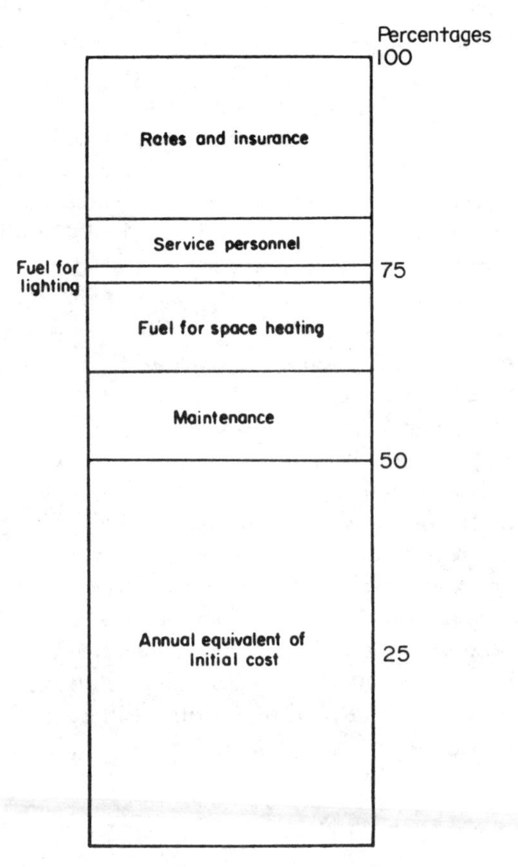

FIG. 17. 1 Annual equivalent costs of a light factory building.

VALUE FOR MONEY

In the commercial sector of the economy, value is something which is objective, since the goods and services are produced not for themselves but for the price they will fetch, and value can be stated in money terms. Thus, in this sector of the economy the value of a building is its value as a factor of production and it can be expressed in money terms. In contrast, the final consumer, whether individual or collective, values goods and services for themselves; the valuation is subjective and no precise money value can be placed upon them. But value consists of not one, but of many attributes, many of which can be looked at as negative costs and hence transferred to the cost side of the value for money balance sheet. For example, if a wall cladding has above average value as thermal insulation, the heating cost saved can be set off against the costs of construction. The concept of cost can be extended to embrace every aspect of cost which arises in the use of the building, whence any saving which follows from employing one design as compared with another will show as a saving in the comparative balance sheets.

While it is difficult to compare bundles of different types of attributes, it is easy to compare sums of money. Hence rational judgements of value for money are simplified by converting as many attributes as possible into money terms. The concept of running costs can be extended to embrace a large range of attributes by measuring the consequences of the design on the costs of operating within the building.(128) For example, it is often possible to measure the additional costs of handling materials resulting from poor layout, or spoilt goods resulting from poor lighting. In this way it is often possible to reduce the attributes of value to little more than appearance and comfort, which can only be evaluated subjectively. The techniques are equally applicable to economic studies of groups of buildings, for example housing estates and even complete towns.(121, 130, 131)

THE TIME FACTOR

The costs of construction, maintenance and servicing are spread in a stream of costs over the life of a building. Leaving aside the exceptional case of a society which does not discount the future and hence has a zero rate of interest, the costs are measured in different units; units payble today and at various times in the future. Future costs are worth less than their face value because they can be secured by investing a smaller sum now, which with interest will grow to the sum required at the date in the future when it is required.

If, for example, a unit of currency is invested at a rate of interest of 5 per cent it will be worth 1·05 at the end of a year, 1·10 at the end of 2 years, and 1·63 at the end of 10 years (Fig. 17.2). In 15 years it will have more than doubled its value, 2·08, and it will be worth eighteen and a half times as much at the end of 60 years. As a result, present monies are worth more than future monies. The comparison of monies payable in the future with monies payable today is made in terms of the present or discounted values. It is these which

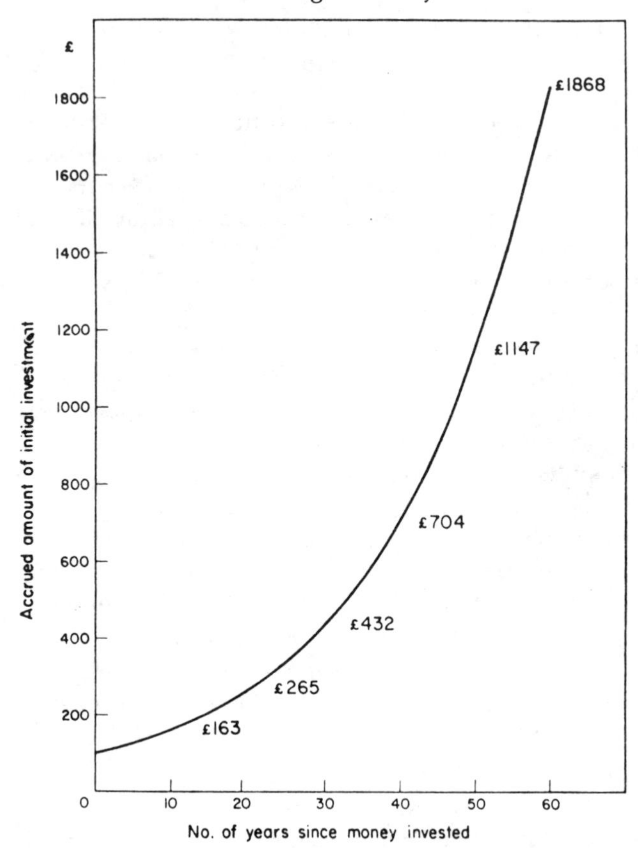

FIG. 17. 2 The amount to which £100 accrues when invested at 5 per cent.

provide the equivalence between present and future values. For example, if a building costs £1000 to erect and needs £500 spent on it in 30 years' time, the total cost of the building in present terms is not £1500 but £1116, that is, the initial cost of £1000 together with the discounted value of the sums payable in the future, in this case 23.1 per cent of £500. It will be clear that the total costs of the building can be met from the sum of £1116, since £1000 of it is used immediately to meet the erection costs and the balance of £116 will be sufficient to meet the bill of £500 in 30 years time if the £116 is invested at 5 per cent in the meantime. Of course, the relationship between present and future money depends on the rate of interest payable in the society for which the discounting is being considered. Summing the discounted values is one way of expressing the "costs-in-use", that is, the equivalent value of all the costs which arise in constructing and using a building.[128] An alternative way of expressing the same relationship is to use the annual equivalent value. This is the sum, payable each year, which has the same value as the present or discounted value. For example, the annual equivalent value of the initial cost is the annual sum which over the period taken will exactly pay off the loan taken to pay the initial cost and accrued interest. It is similar to the annual repayment charged by a building society or other finance house in return for a mortgage on a property. The annual equivalent over a period of 60 years at 5 per cent is 0.0528 per unit. Thus, if the present or discounted value of a stream of payments is £1, the annual equivalent is 5.3*p* (Fig. 17.3).

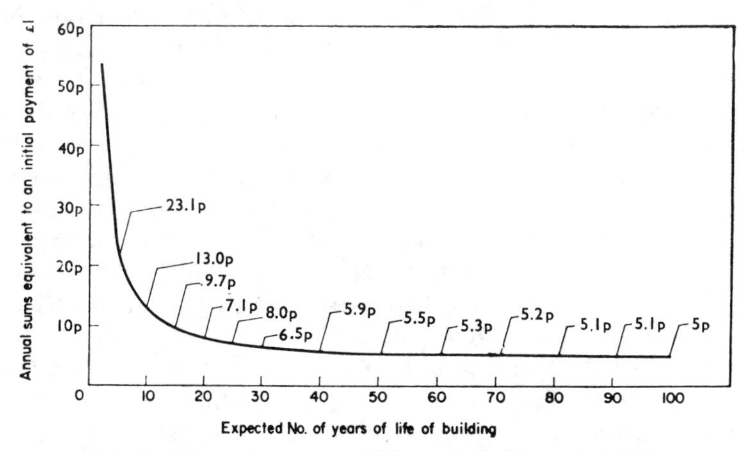

FIG. 17. 3 Annual payments which are equivalent to an initial payment of £1 when interest is at 5 per cent.

Thus there are two convenient ways of expressing the value of a stream of payments: either as the present or discounted value, or as the annual equivalent value. Adding up the face value of the stream of payments is quite meaningless, since this is to add up values which are expressed in different units. Generally, the discounted value has more meaning for the designer since it expresses the stream of costs in terms comparable with the costs of construction. The annual equivalent cost is of meaning to the building user, since it is comparable with a rent or an income.

COSTS-IN-USE

Thus costs-in-use provide a measure of the total or ultimate costs of a building, and it is this cost and not just the initial cost which is to be minimised in relation to the imponderable values of the building. Thus design alternatives can be evaluated by comparing the difference in the costs-in-use as against the differences in appearance and comfort—the two attributes, which, because they are subjective, cannot easily be given a money value. For example, if the difference in the costs-in-use of two buildings is £100,000 but one is drab in appearance and the other attractive and exciting, the £100,000 would be worth spending additionally if the extra satisfaction was valued at more than this sum.

The relationships between the values of sums payable immediately and in the future provide some simple but useful design guides. For example, at a rate of interest at 5 per cent and with a building life of 60 years, *5p* sterling a year is about equivalent to a £1 sterling of first costs. This means that it is not worthwhile to spend more than £1 to reduce the annual costs by *5p,* that is, by a twentieth (Fig. 17.3). For example, it would not pay to spend more than about £100,000 on an automatic device to avoid an annual cost of about £5000 for an operator. On these figures this would set the limit of expenditure on an automatic lift, or an automatic stoker. This indifference level depends, of course, on the rate of interest and on the life of the building. The shorter the life the less it is worthwhile to spend on construction costs to reduce the running costs. For example, at 5 per cent interest it is only worthwhile to spend up to about 50*p* on first costs to save 5*p* on the annual running costs if the life of the building is only to be 14 years (Fig. 16.3). If the life of the building is limited to 10 years the corresponding limit is only about 40*p*. While it would be unusual to erect a building with such a short life, such relationships provide some indication of what expenditure is worthwhile when a building is nearing the end of its life. Thus the shorter the future life of the building, the greater the amount which it is worth spending on future running cost relative to renewal costs, and the less it is worth spending now to save running costs. Hence it is economic to use cheap materials and fittings in short life buildings even if the costs of repairs and renewals is rather high. Similarly, the nearer the end of the expected life of a building the less should be spent on repairs and on components which need replacing.

As explained in Chapter 4 it is the real rates of interest rather than the nominal rates which are of importance. The real rate is approximately the nominal rate less the rate of inflation. Hence if interest is charged at 15 per cent but prices are rising about 10 per cent a year, the real rate of interest is about 5 per cent. During periods of rapid inflation the real rate of interest is often negative. Costs-in-use calculations are made in real terms. Real interest rates are used, that is the expected rate after allowance for inflation. Similarly, real prices are used, that is the prices expected for the period covered by the calculation, discounted for inflationary rises but with allowance for expected changes in real prices. For example, if a costs-in-use calculation includes prices for an appliance, labour and fuel, the price to use would be the current prices adjusted for expected changes in relative prices.

Clearly, the technique of costs-in-use can be applied generally to building designs and to planning solutions. It offers the complete technique required for evaluating designs, since it brings together both the costs of constructing and running the building. The technique is essentially a comparative one. It is concerned not with providing a prediction of the tender price, although the estimating can be carried out so as to provide this as well, but with predicting the relative costs of alternative design features and buildings. Its purpose is to provide a tool for determining the cost consequences of altenative design solutions. The consequences of some design solutions are so obvious that no cost analysis is necessary, but this is by no means generally true. The results of design decisions are often far-reaching; the design of one component interacts with the design of others. For example, the decision to use a two-way rather than a one-way beam system in a roof would affect not only the beam system and columns but also the foundations, the walls, the roof covering and the roof glazing. The effect on costs in a case that was studied would have been to have increased the costs of these things from £31,000 to £39,000; of this difference between £5000 and £6000 would have been running costs.[132]

This type of technique has already been applied to a large range of design problems. For example, cast-iron rainwater goods are generally more expensive than such permanently coated long-life materials as plastics and vitreous enamel ware, because the equivalent annual costs of regular painting are generally greater than the cost of a permanent coating.[129] Glass in walls usually adds to the costs in use of a building because the additional costs of construction, of cleaning the glass, of maintenance and of replacing the additional heat losses, is usually greater than the saving in lighting costs.[132] On the othe hand, the balance of these costs is often in favour of glazing in roofs.[112] The costs-in-use for flat roofs for houses have been found to be 50 per cent higher than for pitched roofs.[133] The life cycle costs of fire protection have also been studied.[134] While the above examples are based on British experience, similar studies have been made in other countries, for example in Australia,[135] in America[136] and Finland.[137] The technique has also been applied to problems in urban planning. The technique can be applied as readily to conversions and renewals of building components and plant as to the design of new construction.[128, 138]

The costs-in-use technique has been given a variety of names varying with the country and context to which it is applied. In America the technique is generally now known as 'life cycle costs'; in engineering the technique is embraced by the term 'terotechnology'; in other contexts and places it is known as 'total costs analysis' or 'ultimate costs'. The technique is related to cost-benefit analysis, cost-effective analysis and threshold analysis. The 'discounted cash flow technique' is an element in all these forms of analysis.

In all cases the approach is similar. It consists of first examining the actual alternative designs, listing the building and other operations to be performed both initially and during the life of the building, eliminating operations that are common to both designs, and pricing those which are different. As far as is possible, every operation and consequences for which the designs differ are priced, so that the total consequential costs of the differences in the designs can be assessed. The differences in the consequential costs are then related to the differences in the appearance and comfort or convenience of the buildings or components.

The costs and benefits to be considered are future costs and benefits, past costs being irrelevant. Thus, for example, the criterion for replacing a component of a building is that the running costs after the change, together with the equivalent annual costs of making the change—the cost of installing the new component—should be less than the running costs if no changes were made at all.[128] If the renewal of the component is not just a technical change but also results in additional benefits, then the value of these benefits should exceed the amount by which the annual equivalent of the installation costs, and the running costs of the renewal exceeds the running costs of the existing component. This type of criterion can be used for examining the economy of putting in a new boiler, or for conversion and adaptation work to an existing building.[128, 139] If, for example, the question is that of putting in a new type of boiler, this will usually offer greater efficiency but will not affect appearance, comfort or convenience other than that which can be expressed in terms of costs. Often the running costs of a boiler will be more than the annual equivalent of the initial costs. In this case, the running costs of the new boiler would need to be less than half the running costs of the existing boiler to make the change worthwhile.

Clearly the analysis of costs-in-use must be based on predictions, and since the future can never be known for certain, prediction errors must be considered. The attempt to analyse all costs and not just initial costs does not introduce the problem of prediction errors; they are present whenever any decision is made which involves a consideration of the future; this is always necessary for buildings since they are durable goods. Any decision about the design of a building necessarily involves assumptions about its future life, its future use, the availability of materials for repairs and renewals, the availability of fuels and about future prices. The costs-in-use technique does not alter this situation, it merely makes the assumptions explicit instead of implicit, so that the effect of the assumptions can be seen and so that it is possible to determine to what extent one design is thought better than another only because of the assumptions made. Clearly, every effort is needed to ensure that the assumptions are realistic. Those about use are particualrly important, for the efficiency with which the design will work depends on the way the building is going to be used and the financial arrangements under which the building user operates. For example, the longer the working day for which the building is used the greater the amount of glazing worth installing, because the amount of heat required normally increases less than proportionally to increases in the working day, and hence is only marginally affected by the extra glazing, while the amount of lighting required increases proportionally to increases in the working day, and hence for at least much of the year full value is obtained for the glazing. Again, for some types of buildings in some countries tax allowances can be claimed on expenditure on running costs at a higher rate than on capital costs, so that the actual costs are much less than the face value of the money spent, particularly on the maintenance and servicing of the building. Such tax discrimination tends to result in lower standards of construction and leads to a waste of the resources of the community. On the other hand, a tax such as VAT which is levied on all maintenance and replacement work, discriminates against adequate levels of maintenance and renewal of buildings.

The main factors, apart from the design and use, which affect the predicted costs are the life of the building and the lives of the components, the rate of interest and the level of prices. While changes in the assumptions may change the absolute costs by considerable amounts, the relative costs may not be much affected. Since the purpose of these cost evaluation techniques is to ascertain which alternative provides the best value for money, absolute costs are less important than realtive costs. For example, if the costs-in-use of two designs are £500 and £400 respectively, and with different assumptions the predicted costs become £600 and £480, the second is still only 80 per cent of the cost of the first and, other things being equal, offers the better value for money. Studies have indicated that, in most cases, the relative costs of a group of designs remains much the same even when the assumptions are changed so as to make large changes in the absolute costs.(128, 140) It would appear that changes in the assumptions only normally change the relative costs when some of the designs being considered have high initial costs and low running costs, while others have low initial costs and high running costs: it is, therefore, necessary to be careful to determine the correct level of assumptions when comparing designs involving the use of such radically different fuels as solid fuel and electricity. Nevertheless, it is important to determine the correct levels of life, interest rates and prices. A conservative attitude will lead to errors. For example, a conservative view on future life will result in an overstatement of costs and will bias the selection in favour of designs with low first costs and high running costs. If a life for the building is taken which is less than the full physical life it is important to make a full allowance for the residual value of the building. A high rate of interest favours designs with low first costs and high running costs, and a low rate of interest the reverse. It is necessary to make an allowance for changes in real price levels. General inflationary changes which affect all prices equally do not need to be considered, although a high rate of inflation will affect the rate of interest; it is the prices which change relatively to one another for which special allowance needs to be made. For example, the price of labour tends to rise more than the price of manufactured goods, so that maintenance work tends to become more expensive as time passes, as compared with new construction. The prices of fuels and certain metals also tend to be variable, and possible changes in their price structure in the future needs consideration. While the uncertainty of the levels of these factors adds to the uncertainty of which design offers the best value for money, it is always present in any decision about design; the use of a formal technique merely makes the uncertainty explicit.

Estimates of the costs of new building work are relatively easy to obtain from priced bills and building price books such as Spons.[141] In Great Britain both initial and running cost data are collected and collated by the Building Cost Information Service.[142] Maintenance and other running cost data are also collected and analysed by the Building Research Establishment [143, 144, 145, 146] and by other departments of government. Generally the resulting estimates are published only for internal purposes.

The costs-in-use is equally applicable to planning problems; for example, to decisions about housing density[147] and to problems about choices as between settlement size and form, including types of transport.[131]

Clearly, the more certain the designer or planner wishes to be about the accuracy and reliability of his choices, the more necessary it is to use these various techniques for evaluating his solutions. The application of such techniques usually requires a knowledge of the technical behaviour of materials, components and engineering services, and of costing techniques which is greater than most designers and planners can hope to have. Usually it will be necessary for the engineers and quantity surveyors to carry out a large part of the design costs evalution, and hence the need for close co-operation between the various professions connected with the design process.

PART V

Natural Resources and Energy in Building

CHAPTER EIGHTEEN

THE FUTURE AVAILABILITY OF NATURAL RESOURCES FOR BUILDING AND THEIR USE

NATURAL RESOURCES USED IN CONSTRUCTION

The major resources used in construction are minerals such as clay, stone, slate, sand, gravel, chalk (for producing cement), and asphalt, metals such as iron, copper, zinc, aluminium and lead, timber directly as wood and indirectly as paper and other products, and oil, coal, and other chemicals for plastic components, paints and other similar based products, and fuels for converting them into building materials and components.

The availability of minerals, used as basic building materials, varies with the locality, region and country. As explained earlier (Chapter 6), the incidence of natural materials varies over the globe. To be of value for building, materials needed to be accessible, and available in relation to each other and to necessary fuel, and must be suitable to the climatic conditions.

In some climatic conditions, for example, where it is hot and dry, stone can be used as dry walling without the need for mortar and clay can be used unbaked. Where there is a great deal of rain, damp-proof walling is necessary.

Bricks cannot be made economically unless there is adequate cheap fuel; the manufacture of cement is also dependent on cheap fuel. In some areas suitable stone for building lies on the surface; sand and gravel can often readily be obtained from the surface. Usually, however, stone, clay, gravel, sand and chalk need to be quarried or at least dug out of the ground. The cavities left become difficult to deal with in densely inhabited and farmed areas. Often the building materials lie under particularly fertile farmland or under land needed for urban development. While with techniques now available it is possible to fill and restore the land, it is nevertheless out of use and unsightly for many years. Such factors reduce the minerals potentially available.

Sand and gravel can, of course, be obtained from the sea-beds around the coasts. This is often of value to areas close to the sea where haulage costs are low, but even so is generally an expensive operation. Supplies are not unlimited, partly because depths are often too great for dredging to be economic and partly because care must be taken to avoid causing coastal erosion. As the supply and price of sand and gravel rises, it becomes worthwhile to use other bulk filling materials for making concrete and concrete products such as clinker, fly-ash, rubble and stone. Some of these materials are more suitable for making thermally efficient building blocks than sand and gravel.

Materials such as clinker, fly-ash, wastes from china preparation, other industrial waste products, and commercial and domestic refuse are all potential sources of raw materials. Many of them can be used for various purposes related to construction, for example, as a filling material for building blocks, for road building and site leveling. Refuse can be used to generate electricity. In the past the cost of utilising waste materials has often been higher than that of alternative materials. Waste materials tend to be bulky and heavy in relation to their value and as a result handling and transport costs may make their use uneconomic, particularly if the place of use is distant from where they are available. Relative costs are likely to change as supplies of materials easier to use become tighter. However, it is generally believed that much more waste material could be used economically.

The development of economic methods of recycling tends to require considerable ingenuity and experiment. Because of the costs of transport techniques are needed to enable waste materials to be used where they are created. This usually involves utilising small quanties by the use of technologies for which the economies of scale are often important. Considerable interest is being shown in developing techniques for recycling waste materials and extracting heat from domestic and industrial wastes at the point of production.

There is a difference of importance between minerals and vegetable materials. The latter can be regenerated, for example, timber reserves can be maintained by replanting. This itself has, of course, a cost. The cycle for regeneration may be long—hardwood forests take several generations to regenerate to an exploitable condition.

In considering the energy used for producing building materials and components, it is necessary to trace the generation and winning of materials as well as their production. Mining and quarrying uses the energy necessary for driving the plant to dig, drill and transport the material before it is processed; energy is also used in manufacturing the plant used in these processes. It can be argued that energy is also used to feed, clothe and house the manpower used in the productive processes.* Energy is also used in raising seedlings,

*However, while price includes a measure of all these uses of energy, it is not possible to use price as a measure of energy because many prices are monopoly prices: currently many fuel prices fall into this category; taxation also enters into fuel prices to an important extent -it is therefore necessary to use shadow prices-energy content is useful for comparative purposes.

planting, in their care and harvesting, but vegetation grows largely by tapping free energy.

A great deal of energy is used in the manufacture of some building materials and components, particularly for some metals. Such materials often have to be fashioned by high energy using processes. Some natural materials such as clay for making fletton bricks (Chapter 6), have a high content of combustible material and only comparatively small quantities of energy are needed to produce them. Materials based directly on a feed stock, such as oil, may not necessarily need more energy in their production than other materials, since only comparatively little energy may be used in processing them. Even in terms of weight, many plastics need less energy than metals for their manufacture. Since plastics are less dense than most other materials they compare even more favourably in energy use in cubic terms (Table 18.1).

TABLE 18.1

ENERGY NEEDED TO PRODUCE SOME COMMON MATERIALS

Product	Oil equivalent by weight* for Feedstock	Conversion	Total	Kilo calories per cc
Aluminium	—	5.6	5.6	158
Steel billet	—	1.0	1.0	82
Tinplate	—	1.25	1.25	102
Copper billet	—	1.2	1.2	112
Glass bottles	—	0.45	0.45	11
Paper & board	—	1.4	1.4	12
Cellulose film	—	4.4	4.4	70
Polystyrene	1.3	1.88	3.18	36
Polyvinyl chloride	0.55	1.4	1.95	28
Low density polyethylene	1.11	1.13	2.24	22
High density polyethylene	1.13	1.2	2.33	24
Polypropylene homopolymer	1.17	1.38	2.55	24

*Tons of oil equivalent per ton of product.

SOURCE: Calculated by I.C.I for U.K. conditions and reported in *Chemical Week*, November 27th, 1974.

The most effective way of saving energy is to reduce the use of materials and labour by achieving the desired ends with less of each. This is achieved over the long term by minimising the costs-in-use (Chapter 17). One method is to restore rather than redevelop whenever this is economic. Many buildings can be restored economically, often with gains to the environment and with social gains. Adaptation often requires more ingenuity than redevelopment, for example changing road alignments without altering levels and widening or building an additional bridge rather than the development of a new one to provide the required capacity. Less energy might be used by restoring existing housing in villages rather than by creating new dwellings in nearby towns, even though additional energy was used for transport.

TIMBER SUPPLY AND DEMAND

Timber is a major material for building, industrial production and as a fuel. While the supply can be regenerated by replanting forests as they are cut down, in the past large areas of timber have been cut down without replacement.

Nearly a half of the world output of timber is used as fuel, mainly in developing countries. Of the balance of timber in world use, two-thirds is used for building and other industrial purposes and a third for paper making.

There is a considerable world trade in timber, many countries being either large scale importers or exporters. America and northern European countries are the main exporters of soft wood and tropical countries export hard wood. Great Britain imports about 90 per cent of its timber needs. It has not in the past replanted its forests and present efforts at regeneration are not thought to be on a large enough scale to produce more than 25 per cent of its future timber needs.[148] Total world consumption is forecast to double by the end of the century and perhaps treble in the next 50 years.[148] While it is thought that supplies might be adequate until the end of the century, the easily exploitable reserves of both soft wood from the northern coniferous belt and hard woods from tropical forests are being exhausted. Future supplies will depend on reafforestation. Clearly it is more expensive to grow timber for the trade than to cut down self-sown forests. The capital cost of planting and costs of tending forests are not released for long periods. With current high rates of interest afforestation tends to be discouraged. The level of timber prices needed to generate the high levels of profits usually required will not materalise until shortages become general. If large-scale replanting is not undertaken soon, timber may become scarce and very expensive unless sufficient suitable substitutes at economic prices become available.

THE USE OF ENERGY FOR BUILDINGS

About a tenth of the gross national product of the United Kingdom is spent on energy; nearly a half of this is used in providing the environmental services for buildings.

Domestic buildings are the major user class of energy for environmental purposes. They use about three-tenths of the total. This is not altogether suprising since they are the major type of building. The domestic sector uses about two-thirds of the energy on space heating, about a fifth on water heating and about a tenth on cooking; little is used for lighting and other purposes.[149] Clearly major savings of energy should be possible in the domestic sector alone; perhaps something of the order of a tenth of national energy consumption.

Industry uses about two-fifths of the gross consumption of primary energy in the United Kingdom.[149] Of course, only part of this consumption is for the heating, ventilating and lighting of industrial buildings; a large proportion is used for industrial production.

Transport uses about three-twentieths of primary energy; other buildings such as offices, commercial buildings, private service buildings and all public sector buildings also use about three-twentieths of primary energy.

Of the primary energy supplied in the United Kingdom, nearly half is oil, over a third coal and about an eighth natural gas; nuclear and hydro power make a very small contribution.[149] The generation of electricity uses about a fifth of the oil, half of the coal and a twentieth of the natural gas. In all, electricity generation uses about a third of the gross energy consumption of the United Kingdom.[149] Not all countries use the same high proportion of energy in the form of electricity as does the United Kingdom.

Whereas the conversion and distribution costs of most fuels are low, so that the consumer receives nearly all the gross energy, these costs are high for electricity and consumers receive only about a quarter of the gross energy input into the generation of electricity. In terms of final net energy obtained by consumers, electricity does not compare as badly as the above figures imply. While the losses on conversion and distribution of electricity are much larger than for other forms of energy, electric appliances use the energy that is delivered to the consumer more efficiently than appliances using other forms of energy. Nevertheless, it would appear in most cases that a better use of energy would be made by using the primary form direct rather than by converting it first to electricity.

The major users of electricity are the domestic sector and manufacturing industry; between them they take three-quarters of the net energy received by final users in the form of electricity. The domestic sector alone takes two-fifths of the electricity generated. However, not only is the domestic sector the largest user of electricity, it also uses a higher proportion for which other forms of energy could be used direct. Over a third of the electricity sold to the domestic sector is used for space heating, although it provides only a fifth of the space heating. Another quarter of the electricity sold to the domestic sector is used for water heating and a further fifth for cooking. Less than a fifth of the electricity used by domestic users is used for purposes for which there is no convenient substitute.

The other users of electricity consume the major proportion for lighting, ventilating and power, for which electricity is the most convenient form of energy. Thus it is in the domestic sector that the greatest opportunities lie for reorientating the forms in which energy is used.

The relative real prices of energy can change considerably over time. For example, in the United Kingdom over the last decade the trend for the prices of electricity have been upwards to a much greater extent than for fuels generally.[150] In fact the price of gas has fallen in real terms. Fuel prices tend to be determined to a considerable extent by political policy. Contracts for imported fuels, mainly oil and gas, tend to be with governments with considerable monopolistic powers. In many countries, notable the UK, fuel and power industries are in the public sector. There is little effective competition; wage, other production costs and even the costs of capital generation can readily be passed on to the consumer, however inefficiently the industry is run. Prices may be raised to meet the need for capital for generation equipment to serve future generations, or prices may be held below costs to assist in holding down general price levels. Future price relatives are therefore uncertain.

Changes in relative fuel costs tend to have only a fairly slow effect on the use of different fuels. This is partly because some fuels are so much more convenient than others for given purposes that prices differences need to be large before it is economic to change but more because most use of energy is in existing buildings (only the equivalent of about 2 per cent of the stock is constructed each year) with plant which is suitable for only one type of fuel. The cost of systems for heating and power are generally too great to justify replacement to allow the use of a different fuel unless relative prices change dramatically and can be expected to maintain the new relationship over a long period. As a result savings in energy costs must be achieved to a much greater extent in existing buildings than new buildings and with existing rather than new plant.

ENERGY SUPPLY AND DEMAND

There are proven oil reserves sufficient for about another thirty years. Proven coal reserves are comparitvely even greater. In both cases proven reserves are only a small proportion of deposits, although for political and economic reasons not all of them are likely to be exploited. But on the one hand, it is necessary to remember that mineral deposits cannot be renewed, they represent the world's capital; on the other hand, new sources and alternatives have been found in the past as previously known resources have been used up: however, the rate of exploitation has never been as rapid as it is at present. Nuclear power at present makes very little contribution to energy. There are large reserves of nuclear fuel and the possibilities of breeder reactors. The major problem with nuclear power resources is that there is no certainty that the extremely dangerous wastes can be safely stored over the extremely long, almost infinite periods, over which they remain toxic. The availability of water power is limited but inexhaustible. The potential tidal power has never been fully utilised, but like water power is inexhaustible. Other possible sources of energy which remain largely untapped are solar and wind power, again inexhaustible, and power from recycling waste. Some energy would, of course, be used in exploiting such sources.

Forecasts for the next four or five decades suggest a moderate decline in the use of oil offset by a slightly greater use of coal and natural gas with the balance made up of nuclear and renewable sources of energy. While there is no shortage of ideas either for synfuels to replace oil and natural gas, or for energy from renewable sources, the substitute sources of energy all appear expensive at current costs relative to existing sources. This would suggest that energy from renewable resources will be marginal. Moreover, the consequences for the environment from the exploitation of renewable sources do not appear all that different in scale from those from traditional sources. However, conventional economic assessment does not adequately value irreplaceable resources relative to replaceable ones. If such adjustments were made, it might be found that some of the replaceable sources of energy could be used economically at the present time.

The relative rises in price of energy, coupled with a decline in the world economy, has reduced demand. Increasing scarcity will tend to raise prices and hence increase the amounts economically worth spending to secure fuel economy. This may reduce demand further relative to what it might otherwise have been. Clearly reducing energy needs is as important as increasing supplies. Both types of action are likely to require additional construction work.

CHAPTER NINETEEN

THE USE OF ENERGY IN BUILDINGS

PURPOSES FOR WHICH ENERGY IS USED IN BUILDINGS

Energy is used in buildings partly to provide environmental services and partly to power the production and servicing carried on in buildings. The main environmental services for which energy is required are heating and ventilation, lighting, the provision of lifts and other internal transport, and cleaning. A great range of functions are carried on in buildings, many of which require energy other than human energy. Within the home, energy is required for water heating and cooking. Functions requiring the use of energy are generally greatest in industrial buildings where energy is required to drive machinery, including internal transport, and for heating and cooling in the course of the industrial processes. Energy is used in considerable quantities in most other types of building, for example, sterilisation and treatment machines in hospitals, computors and copying machines in offices, and refrigeration in shops and warehouses.

Of the environmental services, the main user of energy is heating in cold and temperate countries, and cooling and ventilation in hot countries. As mentioned in the previous chapter, in domestic buildings the production of hot water is also an important user of energy; in such buildings cooking is also important, although to a lesser extent. In buildings with many storeys vertical circulation can also be an important user. The shape and layout of the building, its construction, particularly its cladding and fenestration, and the form in which energy is used all affect the volume of energy required.

The use of energy in connection with buildings is not confined to use within buildings; use of energy about buildings also needs to be considered. This consists not just of lighting around buildings but of energy used generally to create an urban environment, to power the infrastructure and public utility services, and transport between buildings. This embraces the concept of the use of energy in the functioning of urban settlements; the use of energy depending on the type of urban form, that is on the scale, shape, density, arrangement of land uses and forms of communication.[131]

Rather than describe the factors affecting the use of energy in buildings, it is proposed to approach the subject through a consideration of the ways in which energy can be saved in buildings.

WAYS OF SAVING ENERGY IN BUILDINGS

There are ten main ways of saving energy in buildings in providing environmental services:[151]

(1) user action;
(2) siting of buildings;
(3) shape and size;
(4) ventilation;
(5) fenestration;
(6) orientation;
(7) fabric;
(8) type of fuel and form of heating installation;
(9) energy waste;
(10) multiple use of buildings.

Some of the forms of energy saving may imply lower or different standards of consumer comfort and some may only be achieved at a price not worth incurring; the net saving in energy may even be less than the gross saving.

Users can reduce energy consumption by turning-off heat sources which do not add to their comfort, by reducing internal temperatures and by wearing more clothes. The first of these methods implies waste avoidance and not lower standards. Reducing internal temperatures has similar implications only up to a point. The level of clothing worn determines the internal temperatures required for comfort. The more active the occupant, the greater the energy saving possible with a given level of clothing. Even for sedentary work and for resting, the temperature required for comfort is halved as between being nude and dressed in heavy outdoor clothing; a reduction of 14°C.[151] The difference from being lightly dressed, shirt and trousers, and a light business suit is about 3°C. However, extra clothing may in some cases be regarded as inconvenient or uncomfortable.

Tall and exposed buildings use more energy for the same degree of comfort than low and protected buildings. The taller the building in relation to others the more exposed it is. Moreover, the taller the buildings the farther apart and hence exposed they have to be to achieve adequate daylighting with the same amount of glazing,[147] and the greater the turbulence of the air around the buildings—this results in both greater cooling of walls and roof, and if the glazing is not sealed, to greater ventilation losses—if the glazing is sealed, forced ventilation or air-conditioning becomes necessary.[128, 131] Sealed glazing is also likely to be necessary on exposed, noisy or polluted sites. Forced ventilation and airconditioning, particularly the latter, result in the use of much more energy than for heating alone.[128] Again the form of heating and ventilation affects standards of comfort.

The greater the proportion of external surfaces to area and volume the greater tend to be the fabric heat losses. External surfaces are affected by shape, size and form. For small buildings the fabric heat losses can be a fifth less for a square plan than for one with a more extended plan ratio. Similarly fabric heat losses increase with storey height. Again broken forms suffer greater fabric heat losses than agglomerated forms. For example, a detached dwelling has a fabric heat loss of twice as much as an otherwise similar middle terrace house, while an end terrace and semi-detached house has a heat loss about half as much again as a middle terrace house.[149] Of course, the importance of these differences can be much reduced if the fabric insulation values are raised.

Heat requirements and hence energy requirements to replace ventilation losses are reduced proportionally by reducing the rate of ventilation. Beyond a certain point reduction in the number of air changes reduces comfort and increases the risks of condensation.

Of course, the volume of the building and the area of exposed surfaces are reduced by reducing space per occupant, but this eventually reduces standards.

Changes in fenestration and orientation can both reduce summer heat gain, which can be excessive and lead to the need for additional ventilation, and reduce winter heat losses. Glass is a poor insulatant and has a very high U-value. The cost of replacing heat lost through glass often exceeds the value of the light which it transmits.[128] Solar gain is not limited to the summer. Up to a point solar gain in the winter can be of value. The balance of advantage depends on the area of glass in relation to the size of the room. With rooms of traditional size and depth the best balance tends to be obtained with just sufficient glazing for natural daylighting.[139] As the amount of glazing exceeds this level, fabric heat losses and solar gain increase, the one increasing heating costs, while the other increases discomfort and if windows cannot be opened the cost of ventilation—generally there is no gain in useful light. The break-even point for the amount of glazing depends on the design of the space, the way it is used and the hours of use. The heat loss and gain is, of course, broadly proportional to the amount of glass, but varies with orientation—the gain in useful light depends on the position of the glazing in relation to the way the room is used, the hours of use and orientation.[112] Where artificial ventilation is provided, the costs of solar gain depend additionally on the design of the system. Often systems have been designed which are extremely unstable; as a result fluctuations in temperature inside the windows are overcorrected and solar gain can result in additional costs for heating as well as for ventilation. The thermal response of a building will be considered more generally later.

The heat balance of an area of glazing also depends on the orientation and on the location of the building. Both are interrelated. In countries for which much of the time heating needs to be provided , a south-facing elevation is likely to lead to the greatest gains from winter solar heat as well as excessive heating in summer. The fenestration and heating system can usually be designed to provide a net gain. Other orientations will usually not be as favourable, a northern one being the least favourable. The balance will also tend to be less favourable where the position is exposed. On the other hand, where the need is for ventilaiton rather than heating, a northern orientation and exposure to winds, although not to sun, may lead to a better heating balance.

Building materials vary considerably in their value as thermal insulant.(152) Some cladding materials such as glass and asbestos cement sheeting have very poor values. Generally the heavier structural cladding materials such as brick and stone have better insulation values than light materials such as plastic sheets, metal and asbestos cement sheets, and glass. However, the lower the density of the material the better is the insulation value in terms of thickness. Again the greater the thickness the better U-value. Insulation values—U-values—of walls and roof can be improved by adding insulation material to the basic cladding material. Insulation values can also be improved by forming sealed cavities, the wider the better providing that they do not become so wide that the air within them is induced to circulate. Heat losses can also be reduced by using bright surfaced material such as aluminium foil to reflect the heat back.

Improving the U-valve, of course, reduces the gain of solar heat as well as the loss of internal heat. As a result the building becomes more stable in its response to the external climate, temperatures changing less violently than otherwise. Thus ventilation costs may be reduced if forced ventilation or air-conditioning has been installed, as well as internal heating costs.

Minimum values for thermal insulation are often prescribed in building regulations. These, of course, only affect new buildings. Large savings in energy requirements can be achieved by improving the insulation of roofs and walls of existing buildings. Roofs can be given improved insulation by laying fibreglass quilts in the roof space in the case of houses, or fixing some type of insulation board, aluminium backed plasterboard or some other insulant under the roof.(128) The insulation of walls can be improved by adding insulation board or aluminium backed plaster board on battens, or by filling cavity walls.(128) The loss of heat through windows can be reduced by double or treble glazing.

The value of the improved insulation depends on the existing and new

U-values, on internal and external temperatures, and on the way the building is used. Clearly the lower the existing insulation values, the higher the difference between internal and external temperatures, and the greater the hours during which the building is used in the heating season, the greater the gain from improving the insulation values. Whether this gain is obtained in terms of a lower use of energy or higher temperatures depends on the flexibility of the heating system and the way the users control it.

Better insulation is not necessarily cost effective; this depends on the discounted value of energy savings as compared with the cost of the insulation.(128) Improving roof insulation is usually cost effective, wall insulation often is cost effective, but double glazing rarely is cost effective.(128) For a typical dwelling something between a third and two-fifths of the energy currently used might be saved by loft insulation and cavity fill. This would be cost effective for a well-heated house. Double glazing for a dwelling would reduce energy used by about a tenth, but would not be cost effective.(149)

Heating costs are also affected by the type of fuel used and the form of heating system. While electricity is generally used efficiently in electrical heating appliances, although not always in electrical storage heaters, the use of fossil fuels to generate electricity is inefficient; generally only about a quarter of the energy in the fossil fuel is available to the user of the electricity.(149) Thus generally it is much more efficient to use the fossil fuels direct to produce space heating and to heat water. Currently natural gas is the cheapest fuel to use, but oil and coal are both cheaper than electricity for space heating and hot water, even although the installations to use them are more expensive than for electricity.

Efficiency and costs of heating are also affected by the form of heating installation. As explained above, electrical heating appliances, electric fires and even under-the-floor electrical heating are the cheapest type of appliances to install, but their comparatively low initial costs are offset by the high costs of electricity. Gas fired warm air heating systems are cheaper to install than gas fired boilers with hot water radiators. Oil and solid fuel boilers tend to be more expensive than gas fired ones. Pumped hot water systems, while more expensive to install than gravity feed, tend to be more economic to use because of their rapid response rate; they can provide a heating source as flexible as electricity.

In terms of costs-in-use gas fired central heating in dwellings can be as little as only half as costly as electric storage heaters and it can be worthwhile to change from one system to the other in an existing dwelling.(149) Similarly it can be worthwhile to change from electric immersion heaters to gas circulators for heating water and from electric to gas cookers.(149) The extent of the saving can be estimated by calculating the expected costs-in-use of the alternative systems under the anticipated conditions of use.(128)

Possibilities for the future for space heating and hot water include electric heat pumps, district heating, solar heating and aerogenerators. Their economic use depends on their technical development and on their scale of production. Individual solar heating collectors and aerogenerators do not as yet appear to offer an economic alternative in a climate such as that of Great Britain,[149] but solar heating collectors have greater possibilities in countries with more regular and longer hours of sunshine and greater temperatures. The Energy Research and Development Administration of America recently predicted that solar energy will cover a quarter of America's needs by the year 2020 and 7 per cent by 2000. One developer claims that sets of solar heating panels will fall in price by a quarter each time the size of the market doubles, with an eventual fall from $3000 a set to $300.

Clearly energy and costs can be saved by reducing waste such as that caused by inefficient and unsuitable plant, and by unnecessary use. Energy waste also arises from processes where heated materials are allowed to cool before the process is complete and hence subsequently need reheating. Such situations can arise where distance between parts of the plant are large or where there is exposure in inadequately heated buildings or in the open air.

Energy savings can also be achieved by the multi-use of buildings or spaces. This, of course, reduces the need for buildings and economises in their structure and general servicing, as well as in energy. However, if buildings or spaces are used very intensively, additional costs may arise for ventilation and reheating.

The possibilities for energy and cost savings varies with the building, its location and how it is used. Clearly the more thermally efficient a building and its servicing, the less scope for further economies. Location is important not only in terms of orientation but also in terms of the climate to which it is exposed; the higher and more stable the temperature the less heating is likely to be required; the higher the temperature and the greater the humidity, the more likely forced or artificial ventilation will be required. The less a building is used and the greater the density of bodies and other extraneous heat sources, the less heating will be required, although in the latter case more ventilation may be necessary.

ENERGY ECONOMY AND URBAN FORM

Both the form of buildings which are efficient in terms of energy use, including that used in construction, and the need for energy saving for transport between them require a compact but squat form of development.[131] Buildings would need to be erected at their most economic number of storeys. For example, in Great Britain the most economic level is usually one storey for factory buildings, two for residential buildings and four for commercial buildings. In some countries the break-even point is at a greater number of storeys. The buildings would need to be erected reasonably close together both to provide protection to each other and to shorten distances between them. The settlement would need to be kept small, perhaps not more than 10,000 to 50,000 inhabitants. Open space would be located around the built-up areas. The settlements would need to be located in small clusters within which each settlement would be located close to its neighbours.[131]

PART VI

Possible Future Trends

CHAPTER TWENTY

CHANGING CONTEXT FOR CONSTRUCTION AND CONSIDERATION OF FUTURE DEMAND

THE context for construction depends on that for development and on the techniques for construction work. The demand for development depends on a large number of factors the most important of which are population size, structure and distribution, size and condition of the built stock, technology, economic conditions, life styles and the urban forms deemed to satisfy such needs. These factors are, of course, interrelated.

POPULATION TRENDS

The levels of population in Western countries are tending to stability and in some countries declining, while tending to increase in Third World countries although perhaps not as rapidly as in the recent past. The need for development tends to increase with increases in population but the relation between them depends on population structure and distribution as well as on the scale of increases and, of course, on the extent to which need is backed by real demand. Population increase first affects the number of children. The rise in their numbers does not increase the need for dwellings, although it may increase the need for larger ones. It does, however, create an increased need for developments serving childrens' needs such as maternity homes, creches, schools, and factories and other buildings used for producing goods and services for children. As the additional children reach maturity more facilities for adults are required such as dwellings, work places, cultural centres and transport facilities. As the rate of population growth declines households become smaller rather than fewer in number, children decline relative to elderly people, the relation depending on birth and survival rates relating to earlier generations. This is the current situation in most Western countries. Households are tending to get smaller both because there are fewer children in the average family and because the nuclear households are tending to divide into separate households of mature children, working married couples, older people and increasing numbers of divorced. This trend increases the number of small and particularly one-person households, each requiring their own separate dwelling. Associated with such changes are increasing needs for particular goods and services, for example, medical and home care services for the old. These need to be backed by buildings and works for their production.

The geographical distribution of population is no less important. This tends to be changing in opposite directions in Western and Third World countries. In the former population tends to be leaving both the large urban areas, metropolitan areas and large cities, and the villages, and settling in growing medium sized towns. The urban exodus starts in the inner areas with population moving out to the periphery and continues with movement from all areas to new and expanding towns. Movement from the rural areas originates with increasing agricultural productivity and the decline in the number of agricultural workers and continues as rural supplies and services become concentrated in the towns and rural services decline as population declines. In Third World Countries improved mortality has resulted in rural populations greater than local activity can support while industrialisation has created more employment in the towns. Migration to the cities has tended to be larger than they can absorb at the rates of development prevailing. At the same time some towns have declined as their relative advantages for industry have been lost to other locations and as their staple industries have declined.

The migration of population, and industrial and commercial activities creates additional needs for development in the areas to which they migrate. This ranges from additional dwellings, industrial, commercial and social buildings to infrastructure and completely new settlements. The additional needs may not be proportional to the levels of migration as some needs may be satisfied by a more intensive use of built facilities already available. Similarly buildings and other works tend to decline in use less than in proportion to the decline in population because the remaining population tends to make use of the additional space which becomes available. Nevertheless, as population declines so does the need for the available facilities and increasing numbers of units become vacant and gradually decline into dereliction.

URBAN STOCK

The effect of such changes of need on the required level of development and hence on construction work will depend on the state of the stock of urban facilities. Generally the stock of buildings and infrastructure will be older in the areas from which population and activities migrate than in the receiving areas. Its condition will depend on how adequate maintenance has been but generally, even in receiving areas, much of it will be unsuitable for current use and need to be rebuilt or rehabilitated. The growth of the receiving economy will create wealth which will tend to stimulate developments. Hence migration will create a need for development work over and above what would have been necessary had migration not taken place.

The age structure, and standards of maintenance and renewal of urban facilities vary widely even in Western countries. In some countries half of the stock has been developed in the last thirty years, in others only about a third. Often the infrastructure, particularly the underground infrastructure is very old and and poorly maintained. Much new building and rehabilitation is likely to be necessary whether or not migration takes place.

TECHNOLOGICAL CHANGE

Development needs change with changes in technology and life styles; the latter particularly in part on the state of the economy. The pattern of industry changes with changes in tastes, with successful innovations in production and in the development of new products, and with shifts in the advantages in production of one country against others. Increasingly the Third World countries adapt and successfully compete with the Western world in the production of traditional products pushing Western countries to concentrate on exploiting new technologies and sophisticated upmarket goods, and services. Such switches in economic activities requires buildings of different types to those suitable for the production of traditional goods.

Changes in technology, for example, micro-processors, enables goods and particularly services to be provided in new ways and new locations, affecting in particular the economics of scale, communications and transport. Other changes result from changes in the balance and price of energy and technologies for their exploitation. For example, relative shortages and rising prices of oil may limit the expansion of motor vehicle transport, possibly tending to concentrate traffic on more limited routes and activities in denser developments, although at the same time electronic communications are likely to allow more dispersion of activities. A greater exploitation of replaceable forms of energy, such as winds, tides, water power and solar energy (Chapters Seventeen and Eighteen) would create additional demands for development work. Thus technological change is likely both to increase the need for new buildings and works to accommodate them and increase the volume of development needed by increased migration.

Changes in the supply and price of energy is also likely to affect building form. Traditional heavy structures tend to be better insulated against external temperature changes, and for that matter noise, than buildings of a light structure. Glazing is a particular poor insulant (Chapter Seventeen).

Again traditional low buildings tend to make a better use of natural day lighting and ventilation than wide tall buildings and to be less expensive of energy.

Again traditional buildings of heavy construction tend to have a lower failure rate and cost less to maintain than some of the newer forms of buildings. It appears likely that there will be some return to more traditional forms, although more use is likely to be made of prefabricated units and some of the newer innovations in materials and their use. Some of the more recently constructed buildings will need to be replaced (Chapters Six to Nine).

LIFE STYLES

As levels of affluence rise life styles change with a demand for greater comfort, space and ease in servicing. Fashions also change. Together these forces tend to increase the capacity and contents of buildings, for example, equipment, finishes and insulation, and the frequency with which changes are desired. This leads to an increasing demand for new building and rehabilitation both of buildings and some types of infrastructure.

THE ECONOMY

The extent to which such building needs are met depends on the state of the economy both in the long and short term. Few countries appear to have been able to maintain the rates of economic growth achieved in the 1950s and 1960s. Affluence in terms of output per head has risen less rapidly and in some countries has fallen. Where the bargining power of labour has tended to be high, as in Great Britain, earnings of those in work have been maintained in real terms with a fall in employment and in the real earnings of capital. Inflation has been high and real rates of interest have fallen. As a result the market for development has been distorted. Entrepreneurs, except property developers, have had little incentive to invest. The public sector has reduced capital expenditure as the easiest way of reducing public expenditure. The unemployed have little effective demand for construction, while those in work increasingly do building work themselves to avoid the consequences of high direct and indirect taxation (Chapters 4 and 15). When prices were stable governments often followed the Keynes policy of expanding public works, thus creating employment directly and indirectly through the multiplier effect. In situations in which increased employment is followed by wage levels being pushed up beyond the level justified by increases in productivity, increased public expenditure often tends to increase inflation rather than long-term employment. It is, however, necessary to distinguish between public construction work aimed at creating additional facilities and public construction aimed at maintaining existing built environment. Governments, for example, in Great Britain, tend to cut expenditure on buildings and works to amounts quite inadequate to maintain the built stock. This results not only in inadequately maintained housing and other buildings but, less obvious, in inadequately maintained water and sewerage systems, railways, roads and other public utility and transport systems. In the end far more resources will be used as the frequency of emergency repairs and total rebuilding increases.

Life styles and the standards of built environment they set naturally depend on the economy and the standard of living achieved. If there is little increase in income per head, the demand for a better built environment is unlikely to rise. In fact consumer preference may be to sustain and improve the consumption of other goods and services, and spending on built environment may actually fall. On the other hand if income is rising rapidly in real terms there may be more than enough to sustain demands for rising consumption of general goods and services, and demand for built environment may rise, particularly for housing, transport, cultural and leisure purposes. There may be increased demand for the services of public utilities without the awareness of the need to replace existing systems nearing the end of their lives or not really adequate to meet new demands. Consumers, who have been enjoying services at prices which do not reflect replacement costs are generally loath to meet higher charges, especially to cover a backlog of maintenance and replacement. Clearly it is not posible to generalise about the future patterns of demand for development work; these will depend on the way the various trends interact. Future patterns can be projected for each country or region by projecting past trends modified by expected future changes. Clearly there must be some uncertainty about future trends, these will be greater the further ahead projections are pushed. While the future cannot be accurately projected, it is easier to prepare future strategies and avoid costly mistakes if they are prepared in the context of the fullest information (Chapter Twenty-one).

RELATIVE IMPORTANCE OF FACTORS

Which are the more important trends will depend on their relative movements in each country. In Western countries trends in the scale and structure of population are likely to be too small in the short and medium term to create much uncertainty. Population trends in Third World countries are relatively much larger and volatile; their projection is far less certain. Migration is large enough to be important for the future even in Western countries, although the trends appear to have been fairly consistent. There must be some doubt as to how far government policies could modify migration trends. Projections of technological change are generally feasible in the short to medium term because the innovations likely to be exploited have already occurred, although the speed of exploitation must remain uncertain. There must be considerable uncertainty as to future economic conditions. In particular to what extent economies can re-establish steady growth with full employment and price stability or whether they only succeed in achieving a stable balance at a level much below the potential of the national resources available, and the extent and amplitude of trade cycles. There must be considerable uncertainty as to how far Third World countries will achieve a full use of their resources, and how their economies will be integrated with those of the Western world. Trends in life style depend in part on social and moral development, and in part on national economic welfare. The condition and potential of the existing built environment is clearly knowable but adequate measures are not always available. The trends in demand for development work will also be affected by the strategies for development which are followed. These will affect economic trends and life styles as well.

POSSIBLE SCALES OF CHANGE

This is not the place to develop projections for construction work; this would require a book on its own. It is, however, useful to have some appreciation of the scale of changes in construction demand which might materialise in one country.

For Great Britain as a whole it appears unlikely that, in the medium term, say to the end of the century, construction demand is likely to return to the levels experienced in the early seventies. Demand might return to previous levels if the economy was returned to previous levels of growth, and the public and governments raised their levels of appreciation of built environment and its current state.

The number of dwellings required to meet the needs of households over the rest of the century appears unlikely to greatly exceed the present number. However, many dwellings are currently much below acceptable standards, not of the size likely to be required and probably not located where they are likely to be needed. Consequently many new dwellings are likely to be required to replace existing ones; the development of dwellings may not fall below half to two-thirds of the long-term rate of construction and might be higher than current rates. Clearly future demand will depend on many factors including migration and economic trends, and future life styles.

While more primary and secondary school places have been built since 1945 than appear likely to be needed over the next two decades, a large proportion have been built in areas of population expansion, leaving many in old and unsuitable buildings in the metropolitan areas and large cities. New development is likely to be necessary both to replace obsolete buildings and to create additional ones in areas receiving migrants. Nevertheless it appears unlikely that school building will be necessary at anything like past rates.

Although employment in industry is likely to continue to decline over past levels as productivity increases and employment switches to service industry, the combined effects of changing technology, shifts of manufacturing to upmarket and new products, and migration to new areas appears likely to increase the long-term rate of development of industrial buildings.

Despite the shifts to service industry, technological innovation and new locations, it appears unlikely that the high development rates for commercial property in the recent past can be sustained over the rest of the century. Changes in technology appear likely to reduce the rate at which additional space is required, while generally existing commercial buildings are more adaptable than industrial ones.

Future requirements for health and other residential institutions are likely to depend on the extent to which residential care is replaced by more home care, and by a more intensive use of residential space. Achievement in these direcitons would be likely to reduce the need for new buildings below recent building trends.

The need for new public utility networks and terminal facilities will depend partly on increasing consumption and partly on the extent of migration. Public utilities (and road networks) have been constructed as required by housing and other development, mainly in the last century and a half. While there have been extensions and some new trunk systems have been laid, generally the networks have not been renewed since they were built; maintenance has generally been minimal. There is considerable doubt whether renewal of the older parts of the networks, mainly in the cities, can be postponed much longer. Their renewal would considerably increase the future rate of such construction work.

Similarly the need for new terminals and networks for energy services will depend on increasing consumption, migration and renewal. Renewal will arise not only from physical obsolecence but from any changes to new forms of energy source, particularly from exhaustable to replaceable sources of energy. While many electric generation plants and networks are fairly old, oil pipelines are fairly new and gas pipelines are already being replaced. On balance it would appear that construction trends may rise.

Transport has been changing for many decades from rail and water to road and air. The networks of both rail and water transport have been reduced and despite some electrification of the former only very limited renewal and maintenance has been undertaken. Changes in shipping have necessitated the development of new types of port facilities; older ports which have not been redeveloped have declined. While motorways have been constructed, although generally not to the standards necessitated by current levels and weights of traffic, and some improvements have been made to other trunk roads, generally roads have not been improved or renewed relatively to the growth of road traffic. Growth in air traffic also creates a need for the development and renewal of airport facilities. A channel tunnel or even combined tunnel and bridge is a possibility. As a result a considerable increase in such development work over the next two decades appears not unlikely.

Housing has not been maintained to an adequate standard. Moreover, many dwellings have approached or are approaching the stage at which large components such as roofs, external joinery, electric and plumbing services need to be renewed. Large-scale renewal and maintenance work is likely to be necessary if much of the older housing is to continue to give a useful life. Given adequate finance such work could increase considerably in the future.

While some types of building have been maintained to a more adequate standard than housing, the maintenance of infrastructure has, as indicated above, generally been far from adequate. A substantial rise in non-housing maintenance is necessary if the facilities are to function adequately.

As indicated earlier, the future levels of construction activity depend on a large number of factors. While it appears likely that future levels may be lower than past trends, the types of construction work appear likely to change with increases in maintenance relative to new work and of engineering work relative to building work.

CONSEQUENCES OF CHANGE

Such changes in the demand for the services of the industry would have a profound effect on its organisation, on the output of the various parts of the industry such as manufacturers of materials and components, contractors and professional advisers. Unless the industry was reorganised to meet the changed pattern of demands some resources would inevitably be wasted, while others would be in short supply. Such reorganisation would take several years and could only be adequate if future markets were known several years ahead. While such knowledge cannot ever be certain, a knowledge of probable markets is of value. As indicated earlier the future market depends on many factors some of which can be projected from a study of trends in the various fields affecting the market and some of which depend on government policy. Government is concerned in part with using resources in the best way to meet the asperations of the community and also needs to understand the ways in which the economic and social climate is likely to develop, the options open and the best strategies to follow. These matters will be discussed in the next chapter.

CHAPTER TWENTY-ONE

POSSIBLE CONSEQUENCES OF FUTURE DEMAND ON THE CONSTRUCTION INDUSTRY

AS indicated in the previous chapter there are a number of factors which affect the future demand for construction work. Broadly these can be divided into two groups,; firstly factors which relate to the national climate, to situations which arise from past trends at home and abroad, and to collective national decisions which arise from them, and secondly to factors related to public policy. The first type of factor includes the growth, structure and location of population, growth and pattern of trade, changes in living styles and consumer preferences. The second type of factor includes government policy for finance and taxation, planning and town development, industry and communications.

GOVERNMENT POLICY AND CONSEQUENCES

The government, central and local, together with the nationalised industries are directly responsible in this country for about half the demands on the construction industry as well as indirectly influencing private development. As argued earlier (Chapters Five, Twelve and Twenty) governments have tended to give little attention to the consequences of their policies on the built environment and the construction industry, or even to the long-term needs of the environment. If governments were to recognise the importance of the built environment and the need for long-term planning of town and country development, and determined and published long-term policies, the uncertainties of projecting future demand for construction would be greatly reduced and the industry could shape itself in some certainty of the demands upon it.

As explained in the previous chapter the starting point for a consideration of government policy for the built environment is the projection of the future contexts. A study of these provides a basis for determining the issues to be considered. While government policies may affect the level and structure of population, their influence on location is far more direct. Negatively they can reduce development and redevelopment in particular locations by such policies as withholding planning consents and not developing necessary infrastructure. Positively their powers are more limited unless they finance industry and direct labour. The provision of planning permissions and the development of infrastructure, while encouraging new development or redevelopment will not necessarily attract people and industry to a location which does not appeal to them. If grants and subsidies are sufficient firms may set-up in a location which is otherwise considered unfavourable but usually the national resources will be used less effectively than they would otherwise have been used.

Governments need to have a strategy for population location if only as a basis for policies for developing the built environment. There are many options to be considered. For example, in the case of existing settlements possible options include aiming at stabilising the population or physical size, or expanding or reducing either of them. Built environment, both buildings and infrastructure, is subject to deteriation and obsolesence. Redevelopment is necessary to offset physical decay and to meet changing needs even when an area is planned to contract. Because redevelopment tends to be expensive relative to new construction on virgin sites, private agencies tend only to undertake redevelopment when either greater densities can be achieved or upmarket uses are possible and values are created sufficient to cover the additional costs and provide a return on capital. Otherwise private developers tend to concentrate on developing virgin land at the periphery of settlements. As a result obsolete and physically worn-out residential, commercial and industrial areas tend to become unoccupied and derelict. Public utility organisations, generally part of the public sector, are also reluctant to redevelop existing services or even undertake adequate maintenance as long as the services function at an acceptable level because of the consequences for their prices.

Deterioration and obsolesence tends to be patchy so that it is seldom worthwhile to redevelop areas as a whole. Gradual redevelopment in rundown areas tends to be uneconomic because it is difficult to attract sufficient users to pay for the services. On the other hand, there are usually adequate useful facilities in rundown areas which are duplicated when new areas are developed. Moreover, changing location raises difficulties in servicing and administration, and their real costs, and results in social disturbance and hardship. Such external costs are not taken into consideration by developers. Thus, while private costs of redevelopment of rundown areas are greater than costs of development on virgin land, because of external costs community costs are often lower and there are therefore grounds for public intervenion either directly or through subsidies. This represents a range of options for governments to consider.

Population and industry moves from other locations, for example, small towns, villages and the countryside because services are no longer adequate. Governments can modify such migration in a free society only by changing the balance of economic and social advantages by improving services through policies for siting schools, hospitals and other public services, improving communications and so on, or through subsidies. If such strategies are to be successful a balance is necessary to encourage both residents and jobs. Stabilising settlement patterns and achieving growth against the preferences of the community may not be economic, even when external costs are taken into account and may be attempted for political or defence purposes.

Built development has a long life; most developments take several years to complete; maintenance cycles are of several years' duration. As a result development decisions have long-term consequences, while development and maintenance strategies need to be sustained over long periods. In contrast governments tend to take short-term views related to periods of power. Such a philosophy fits badly with development needs and results in a waste of national resources. Long-term policies need to be sustained for the construction and maintenance of the built environment both to avoid waste from locking up resources in incomplete developments and from a failure to maintain facilities to an adequate standard to avoid deterioration and dereliction. At the same time building construction resources are wasted by fluctuating loads of work.

While changes in economic conditions often make it difficult to sustain regular building programmes, the factors behind planning strategies tend to change fairly slowly and predictably so that sustained policies should be possible. It should be possible for developers to plan well ahead on the basis of predictable levels of density, building form, settlement design and scale, availability of lines of communication, public utilities and so on.

Development strategies cannot be implemented without appropriate financial, fiscal and administrative resources,or without an adequate supply of building resources such as labour, materials, plant and management. All such resources need to be kept in balance with the development programme. For example, in the case of private housing development implementation of a programme depends on such factors as adequate loan funds to builders and mortgage funds for purchasers at acceptable rates of interest, with incomes, taxes and subsidies providing sufficient resources and incentives to would-be purchasers, together with land with planning consents and public services, and sufficient building resources in the areas of development.

While it is self-evident that development can only take place when all the factors are appropriate, frequently governments pursue incompatable policies with the result that their aims are not achieved. Whatever the planning and development strategy it will only be implemented if all the necessary conditions are created in a balance way. The most efficient use of construction and other resources will be achieved if the programme is organised so that the demand for resources is either uniform or changes in ways consistent with the creation and rundown of construction and other facilities. The cycles for most facilities are long, taking several years to create and many more years to be fully exploited. For example, it takes several years to train both professional and skilled operatives and only a small proportion can be utilised in other fields without several years retraining if demand declines. There is similarly a long embryonic period for building factories to produce building materials and plant, and little use for them or their products if construction declines.

CONSEQUENCES FOR DESIGNERS

Changes in population scale and structure, technology, economic affluence and living styles, and other factors have consequences for building design and settlement development. They have an effect on such design factors as the size and layout of building, their form, construction and equipment, and the size, shape, form and density of settlements.

Generally only the largest building clients, mainly in the public sector, carry out adequate research as a basis of design and planning strategy. Evidence suggests that this has not always been very thorough or far sighted. Fashion often appears to be a stronger influence than scientific analysis. Close attention to trends in user needs, in technology and the other factors discussed in this and the previous chapter could have a substantial effect on the economics of building and on the nature of building demand.

The design side of the industry is fragmented between many design offices and professions. Many of the design offices are very small, while large ones tend to be divided into largely autonomous design groups. Professional fragmentation increases the difficulty of applying all the design skills at the best stage of the design process. Fragmentation into many autonomous design group increases the difficulty of passing on knowledge and experience. Both are increasingly more important as technological development increases the range of solutions and the need for technical knowledge. Design research is generally very limited. Its fragmentation increases the difficulties of acquiring an adequate knowledge of its results.

Most buildings and works tend to be designed from first principles, although many of the problems to be solved are similar. In the past most buildings were constructed on the basis of traditional solutions or pattern books. Despite the number of separate designs groups the differences in design for many classes of building are small. Better value for money might be obtained if there were fewer but better researched designs for each class of building. This would involve fewer but larger and multi-disciplined teams.

Changes in the technological content of building and construction is likely to increase the need for technological specialisation with fewer outlets for general design skills. The skills of designers are also likely to need to change with changes in the relative volume of structural, civil, mechanical and electrical engineering and with changes in the relative importance of rehabilitation and maintenance work relative to new construction.

Similarly there may be changes in the relative importance of operative as compared with administrative and control professional work both for buildings and for planning. Control of planning, design and building construction tends to be increasing both for safety and aesthetic reasons. Large numbers of professional people are employed in regulation work including planning control, tree and landscape preservation and the administration of fire and building regulations. The best balance of skills required for the different functions may not be the same.

The training of bulding professionals tends to be a long process lasting several years. Specialisation might improve performance and reduce training periods but would reduce flexibility and make it even more important to assess the relative needs for various types of skills several decades ahead.

CONSEQUENCES FOR CONTRACTORS

Similarly the demand for building and construction contractors depends on the scale and types of work to be performed. Generally only large contractors can handle large contracts and can operate over large geographical areas. The smaller and more specialised firms are likely to be affected by both changes in the type of work and its location. Even large firms might need to be reconstructed to meet changes in demand. Firms of all sizes and types are needed to efficiently meet the range of likely demands. The type of organisation and plant requirements vary with the type of work. Thus contracting firms need to match the scales and types of demand and to be able to evolve without wasting resources as demand changes. Thus a knowledge of likely future demands is necessary if the industry is to be efficient.

Again the contractors' demand for building skills, operative, professional and managerial depends on the types and scales of demand. Acquiring skills takes several years and it is necessary to have a firm idea of future needs if shortages and excesses of skills are to be avoided with their effects on costs and social hardship.

CONSEQUENCES FOR MANUFACTURERS

Equally the demand for building materials, equipment and construction plant depends on the scale, structure and location of building demand.

Changes in the scale of development, in forms of construction, standards, and in the balance of new, rehabilitation and maintenance work, and in building and engineering work all affect the demand for manufacturers' products. This affects both manufacturers and the distributors. Since again the time taken to bring new production on stream is several years, with a useful output potential for many years ahead, plant planning needs to be based on medium term demand forecasts.

Thus efficiency in every aspect of building requires a steadily evolving development of demand and reliable forcasts.

CONCLUSIONS

THE achievement of economy in building and construction is of considerable importance to the industry, to the clients and to the community. There is no single way of achieving economy in construction; success depends on the rationalisation of every aspect of design, production and organisation. The creation of buildings is more complex than the creation of most other goods, both from a technical and from an organisational point of view. Local production factors are probably both more variable and more important in the construction industry than in most other industries.

Traditional forms of building vary considerably from one country to another according to the local supplies of labour, materials and managerial ability. But traditional building is constantly changing as innovations in materials and techniques are developed. There are few radical differences between traditional and non-traditional forms of building, partly because the newer forms of construction grow out of past experience, and partly because traditional building absorbs the worthwhile innovations as they are developed. The pace of development of innovations in building tends to be gradual and really radical innovations are seldom realised. The absence of dramatic changes in building construction as compared with manufacturing industries is a result of a multiplicity of factors; the long period over which the industry has developed, the difficulties of mechanising assembly as compared with manufacturing work, the complex nature and range of the final products, and the absence of any great growth potential in the market.

The awareness of the need for increasing the efficiency in the industry is perhaps greater now than ever before and greater interest is shown in developing innovations. So far, the display of technical ingenuity has been greater than the achievements of economic success. The range of materials and components available to the industry is increasing, but because of the difficulty of competing with the cheap traditional materials innovation does not necessarily lower costs. Similarly, new building systems have added to the range of building solutions but have not yet achieved economies of construction. It is not yet certain whether there are any potential economies of importance to be obtained along these lines. No doubt any successful innovations developed in this field will be absorbed in the field of traditional building, as has occurred in the past.

The rationalisation of building construction can and does proceed in two ways; by evolution and by revolution. The evolutionary methods consist of rationalising the many aspects of traditional construction, introducing prefabrication, mechanisation and new methods of organisation where they are likely to achieve a saving in resources. The revolutionary methods consist of developing ways of exploiting new materials and techniques. In the short run, such methods provide a less certain way of achieving economies but they provide the most likely way of obtaining large savings.

It is unlikely however that there is any simple single way of obtaining substantial reductions in the resources needed to build. Economy is more likely to be achieved as a result of developments in all the aspects of building. Standardisation and dimensional co-ordination of materials and components should bring economies in their production and use. Economies can be obained from a better exploitation of existing materials and from the use of new materials. There are few solutions which are equally applicable under all situations; the best solution depends on the local conditions to be satisfied and on the local supply and costs of labour and materials.

While mechanisation has already resulted in substantial economies in some directions, there is a limit to the potential of mechanisation in an assembly industry, and much of the potential has already been exploited, at least in the more advanced countries. The potential for mechanisation is increased when *in situ* work is replaced by factory-made units, but frequently this demands a substitution of more expensive materials and does not always produce any overall economy. In fact, the key to economy in construction work probably lies in organisation.

Whatever the form of construction used, the organisation of the job is of considerable importance. There is no standard solution; each building job requires careful analysis and the application of the best principles of planning and programming. The development of new techniques can supplement but not replace the managerial skill necessary for the planning and organisation of the individual job and for the management of labour. The differences in the levels of efficiency obtained by different contracting organisations are considerable. A general improvement to the levels of efficiency achieved by the best organisations would probably result in a greater measure of economy than can be achieved by innovations in technique.

Similarly, the efficiency of individual designs is very variable. Again, economy probably depends more on the appliction of personal skill and knowedge to individual design problems than on innovations in construction techniques.

Institutional arrangements within the construction industry tend to impede economy in the use of resources, partly by separating the various parts of the construction process, and partly by creating conflicts between the interests of the various parts of the industry. While the difficulties created by the division between design and construction are important, the balance of advantage does not appear to lie with any of the forms of unification so far in use. Similarly, the direct and indirect advantages of competitive tendering appear to outweigh the disadvantages. Without a fair amount of work being let by competitive tendering there would be no basis on which to negotiate, and in the absence of some type of priced bills there would be no basis for cost planning.

It is not certain to what extent limitations arising from the organisation of the contracting industry result in serious losses of efficiency. Moreover, the economies possible from a rational organisation of the job are unlikely to be fully realised where the responsibility for organisation is divided, or where the size of the jobs is very small. Furthermore, rationalisation of production is unlikely to be economic for very small production organisations. On the other, size is no guarantee of efficiency; the greater the size of the organisation the greater the need for a system of built-in incentives to efficiency, and the more important is a sensitive system of costing.

Building economy does not stop with the erection of the building. Buildings are very durable and design economy must take into consideration the costs of using the buildings throughout their life. Far too little is yet known of the way the design of different components of a building interact one with another and with the ultimate costs.

The total costs of a building are about double the initial costs. Maintenance and rehabilitation tends to be labour intensive and its costs tend to rise faster than initial costs. Few buildings will satisfy needs over their physical life; most will need to be converted or demolished during the span of possible physical life. Often it will be economic to build in flexibility and adaptability to ease and reduce the costs of likely future changes to meet new requirements. A large proportion of the running costs of buildings arises from the use of energy, the cost of which is likely to rise as existing sourcs are exhausted. Substantial savings in the cost of energy can be achieved through vigorous analysis at the design stage and subsequently by incorporating thermal insulation, reducing air changes and rationalising methods of heating.

Building is a complex activity which uses a large proportion of the resources of most nations; economic considerations tend to be dominant in determing forms of construction. In such an activity there can be no simple, single way of achieving efficiency. The overall efficiency depends on the efficiency of the individual parts. Economy in building must, therefore, be sought through efficiency in the design and the production of the materials and the components, in the methods of assembling them, in the design and construction of individual buildings, in the organisation of the industry, and in the organisation of individual firms and individual jobs.

The demand for built construction depends on many factors, of which changes in population scale, structure and location, technology, life styles and the economy in relation to the urban stock are amongst the most important. Changes in demand for the services of the industry can have a considerable effect on its organisation, output and level of employment, and hence on the demand for the various factors of production. If substantial changes in demand are to be met without wasting resources, it is necessary to have an understanding, some years ahead of the changes, of what the requirements are likely to be, so that government, clients, material producers, contractors and professionals, training organisations and so on have sufficient time to develop and implement appropriate policies to meet the expected changes.

REFERENCES

1. SAMUELSON P. A., Economics: An Inroductory analysis, McGraw-Hill
2. MARSHALL A., Principles of Economics, Macmillan
3. HILLEBRANDT P. M., Economic Theory and the Construction Industry, Macmillan, 1974
4. *Engineering News Record (U.S.A.),* 21 March 1963. *Statistical Abstract of U.S.A.,* 1960.
5. *Mechanical and Industrial Building in Sweden,* Byggnadsfackens Utredningssavd (unpublished).
6. *Chartered Surveyor* (Various dates). *Wages and Income since* 1860, Bowley. *National Income and Expenditure,* H.M.S.O. (various dates).
7. *Government Policies and the Cost of Building,* United Nations, 1959.
8. *Nationwide Building Society* Occasional Bulletin No. 135, 1976
9. MAIWALD, K., An index of building costs in the U.K., *Economic History Review,* 2nd ser., v. 8 (2).
10. JAGGARD, W. R., *Experimental Cottages,* H.M.S.O., 1921.
11. LEGGET, R. F. and HUTCHEON, N. B., Building trends, *2nd C.I.B. Congress,* 1962.
12. TAKEYAMA, K., Design and construction in Japan and in some other Asiatic countries, *2nd C.I.B. Congress,* 1962.
13. WOODHOUSE, W. M., Developments in the local building industries of the Commonwealth, *J. Roy. Soc. Arts* **110** (5073), 1962.
14. *Survey of Building in Sandstone in Scotland,* N.B.S. Special Report No. 20, H.M.S.O., 1953.
15. *A Work Study in Blocklaying,* N.B.S. Technical Paper No. 1, H.M.S.O., 1948.
16. STONE, P. A., International comparison of building costs with particular reference to U.S.A. and G.B., *Bull. Oxford Inst. Statis.* **22** (2)
17. FREEMAN, I. L., Comparative Studies of the Construction Industries in Great Britain and North America, CP 5/81 Building Research Establishment, 1981.
18. FLEMMING, M. C., Structure, output and prices of the building industry in Australia, *The Building Economist* (2), June 1962.
19. RIDGE, M. J., The assessment of new developments in the Australian gypsum plaster industry, *2nd C.I.B. Congress,* 1962.
20. BOWLEY, M., *Innovations in Building Materials,* Duckworth, 1960.
21. *Cost, Repetition, Maintenance, Related Aspects of Building Prices,* Economic Commission for Europe, U.N., 1963
22. *Economist* 2 May 1981
23. *Cost Saving through Standardisation, Simplification, Specialisation in the Building Industry,* O.E.E.C., 1954.
24. BISHOP, D, Dimensional Tolerances and the development of building systems, C.S.8, Building Research Station, 1963.
25. HARRISON, M., A century of prefabrication, *The Builder,* 21 June 1963.
26 BISHOP, D., Large panel construction, *Architect and Building News,* 5 February 1964.
27. *House Construction, Post-war Building Studies No. 1,* H.M.S.O., 1944.
    Private Enterprise Housing, Ministry of Health, H.M.S.O., 1944.
    Report of the Committee on Scottish Building Costs, Dept. of Health for Scotland, H.M.S.O., 1948.
28. WEAVER, SIR L., Cottages—their planning, design and materials, *Country Life,* 1926.
29. FREEMANTLE, F. E., *The Housing of the Nation,* Allen, 1927.
30. ZAIMAN, A. and MCINTYRE, W. A., *Economic and Manufacturing Aspects of the Building Brick Industries,* Building Research, Special Report No. 20, H.M.S.O., 1933.

31. *House Construction,* Post-war Building Studies No. 23, H.M.S.O., 1946.
32. *House Construction,* Post-war Building Studies No. 25, H.M.S.O., 1948.
33. *The Cost of House-building,* First Report, Ministry of Health, H.M.S.O., 1948.
34. *Hansard,* House of Commons, H.M.S.O., 6 April 1948.
35. *Housing Returns,* Cmd. 9681 and Cmd. 9366, H.M.S.O., 1956.
36. COX, O. J., Factory production in housing, *Contract Journal,* 12 April 1962.
37. *Economist,* April 29 1980
38. *Temporary Housing Programme,* Ministry of Works, Cmd. 6686, H.M.S.O., 1945.
39. *Temporary Accommodation,* Ministry of Health and Ministry of Works, H.M.S.O., 1944.
40. *Temporary Housing Programme,* Ministry of Works, Cmd. 7304, H.M.S.O., 1948.
41. BIRD, E., Industrial housing, *The Builder,* 1 February 1963.
42. SHIFFER, E. T., The machine is the new workman, *Interbuild* **9,** October 1962.
43. *Building in Russia,* Eastern Federation of Building Trades Employers, 1962.
44. DUPRAT, N. J., *Annales de l'Institut Technique du Bâtiment et des Traveaux Publiés,* October 1962.
45. *Hansard* (Col. 1311), House of Commons, H.M.S.O., 17 March 1964.
46. MADIGAN, J. J., Letter to *The Builder,* March 1963.
47. CHASTEN,D D. M., Building practises in America and Russia, *National Builder,* June 1963.
48. PERKIN, G., Industrialised building on the Continent, *The House Builder,* November 1962.
49. CRAIG, C. N., A battery method for site casting of internal walls and floor panels, *Architect and Building News,* 3 July 1963.
50. Jackblock construction, Architect and Building News, 7 February 1962.
51. Lift-slab construction, *The Builder,* 17 January 1964.
52. The Cauvet system, *The House Builder,* January 1963.
53. Industrialised building, *Municipal Journal,* 25 January 1963.
54. IREDALE, R., The Nenk method in operation, *Architects Journal,* 13 March 1963.
55. STERN, E. G., The technology of building with timber, *B.A.T.A. (Australia),* January 1963.
56. *The Construction Industry,* National Economic Development Council, H.M.S.O., 1964.
57. *New Methods of House Construction,* N.B.S. Nos. 4 and 10, H.M.S.O., 1948-9.
58. *A study of Alternative Methods of House Construction,* N.B.S. Special Report No. 30, H.M.S.O., 1959.
59. TURIN, A. D., Report of Seminar at the A.A., *The Builder,* 29 March 1963.
60. REINERS, W. J. and BISHOP, D., Construction of multi-storey flats, *The Builder,* 27 April 1962.
61. BARR, A. W. C., Housing from the factory, *J. Soc. Housing Manager,* January 1963.
62. LEON, G., The Economics and Management of System Construction, Longman, 1971.
63. Industrialised building, *The Economist,* 14 July 1962.
64. BISHOP, D., *Systems of Construction—Economic Performance,* Building Research Station, 1963.
65. WOOD, K. M., Industrialised building, *The Builder,* 24 May 1963.
66. LEMESSANY, J. and CLAPP, M. A., Resource Inputs to Cosntruction CP76/78, Building Research Establishment, 1978.
67. *The Story of Clasp,* Ministry of Education Building Bulletin No. 19, H.M.S.O., 1961.
68. STONE, P. A., A survey of the annual costs of contractors' mechanical plant, *J. Ind. Econ.* 4 (2), 1956.
69. STONE, P. A., The output and fuel consumption of mechanical plant, *The Contract Journal,* 10 December 1959.
70. Ready Mixed Concrete, Cmnd 8354 H.M.S.O, 1981
71. EDEN, J. F. and PIPPARD, N. S., The site handling of materials for traditional housebuilding, *The Builder,* 16 November 1951.
72. *Mobile Tower Cranes,* N.B.S. Special Report No. 31, H.M.S.O., 1960.
73. STONE, P. A., The costs and economics of contractors' plant, *The National Builder,* July 1961.
74. CLAPP, R. A., Weather conductons and productivity, CS32, Building Research Station, 1966.
75. *The Cost of Housebuilding,* Second Report, Ministry of Health, H.M.S.O., 1950.
76. *Productivity in Housebuilding,* Second Report, N.B.S. Special Report No. 21, H.M.S.O. 1953.

77. STONE, P. A. and REINERS, W. J., Organisation and efficiency of the housebuilding industry in England and Wales, *J. Ind. Econ.* **2** (2), 1954.
78. BROUGHTON, H. F. and PIPPARD, N. S., Programming of traditional housebuilding, *The Builder,* 11 November 1955.
79. NUTTALL, J. F., Some principles of the production control of building work, *J. Ind. Econ.,* November 1961.
80. PIPPARD, N. S. and BUCKLE, W. G., Prefabrication in traditional housebuilding, *Prefabrication,* May 1954.
81. *Production in Building and Civil Engineering,* Ministry of Works, H.M.S.O., 1945.
82. KELLEY, J. E. and WALKER, M. R., *Critical path planning and scheduling,* Proc. Eastern Joint Computor Conf., 1959.
83. MALCOLM, D. G. and others, Application of a technique for research and development program evaluation, *Operational Research* **7** (5), 1959.
84. NUTTALL, J. F. and JEANES, R. E., The critical path method, *The Builder,* 14 and 23 June 1963.
85. NUTTALL, J. F., *The control of repetitive construction,* RS 34, Building Research Station, 1965.
86. *Organisation of Building Sites,* N.B.S. Special Report 29, H.M.S.O., 1963.
87. EMMERSON, Sir H., *Survey of Problems before the Construction Industries,* Ministry of Works, H.M.S.O., 1963.
88. JEANES, R. E., *The study of operative skills,* CS 30, Building Research Station, 1967.
89. NELSON, JEANES and WARRINGTON, *Building, occupations and training,* CP25/68, Building Research Station, 1968.
90. *Incentives in the building industry,* N.B.S. Special Report No 28, H.M.S.O., 1958.
91. SKOYLES, E. R., *Materials wastage—a missuse of resources,* CP 67/76, Building Research Establishment, 1976.
92. *The Placing and Management of Contracts for Building and Civil Engineering Works,* Ministry of Public Buildings and Works, H.M.S.O., 1964.
93. *The Builder Survey,* The Builder Ltd., 1962.
94. EDEN, J. F. and GREEN, J., The integration of building and engineering design in hospital building, *R.I.B.A. Journal* **70** (7).
95. FORBES, W. S. and SKOYLES, E. R., The operational bill, *Chartered Surveyor,* **95** (8).
96. The functions and uses of bills of quantities, *Chartered Surveyor,* **95** (3)
97. *Organisation and Practices for Building and Civil Engineering,* Ministry of Public Buildings and Works, H.M.S.O., 1964.
98. PORTER, R. W., The Soviet construction industry, *The Builder,* 30 November 1962.
99. ALSOP, K. and DAY, A. G., The French Agreement System, *The Builder,* 26 July 1963.
100. PAIGE, D. and BOMBACH, G., *A Comparison of National Output and Productivity, O.E.E.C., 1959.*
101. Interfirm comparisons for the building industry, *National Builder,* July 1963,
102. COOK, E. J., The builder's use of capital, *The Builder,* 28 June 1963.
103. LEA, E. and LANSLEY, P., Building demand and profitability, *Building,* March 14 and 21, 1975.
104. FINE, B., Tendering strategy, *Aspects of Economics of Construction,* George Godwin, 1975.
105. FORBES, W. S., production cost information, *Aspects of Economics of Construction,* George Godwin, 1975.
106. STONE, P. A., Departmental accounting in the construction industries, *Cost Accountant,* **35** (11), 1957.
107. The Architect and His Office, R.I.B.A., 1962.
108. ALLEN, W., The architectural profession in North America, *The Builder,* 25 May 1963.
109. ROBERTSON, D., Cost information for designers and building owners, *Aspects of Economics of Construction,* George Godwin, 1975.
110. WALES, C. A., Estimating in the United States of America, *Chartered Surveyor,* **88** (10), 1956.
111. HEFORD, J. J. V., *The French Building Industry,* R.I.C.S.;, 1963.
112. *The Economics of Factory Buildings,* F.B.S. No. 12, H.M.S.O. 1962.
113. REINERS, W. J., Cost research, *Chartered Surveyor,* **90** (3), 1957.

114. CRAIG, C. N., Factors affecting economy in multi-storey flat design, *R.I.B.A. Journal,* April 1956.
115. *The story of Post-war School Building,* Ministry of Education, H.M.S.O., 1957.
116. BROUGHTON, H. F. and PIPPARD, N. S., Building and planning—economy in traditional house-building, *Municipal Journal,* 23 March 1950.
117. *Flat and Pitched Roof Construction,* Langley, London.
118. MANNING, P., *The Design of Roofs for Single Storey General Purpose Factories,* Dept. of Building Science, Univ. of Liverpool, 1962.
119. Cost comparison of alternative roofs suitable for local authority type flats, *The Chartered Surveyor* **92** (9), 1960.
120. *Quicker Completion of House Interiors,* Ministry of Housing and Local Government, H.M.S.O., 1953.
121. STONE, P. A., Housing, town development, land and costs, *Estates Gazette,* 1962.
122. KNIGHT, T. L. and DUCK, A. E., the costs of lifts in multi storey flats for local authorities, *Chartered Surveyor* **94** (8), 1962.
123. *Building Research* 1957, D.S.I.R., H.M.S.O., 1958.
124. BELLAMY, A. A., High flats in U.S.A., *Housing Review,* Jan.-Feb. 1958.
125. Factors affecting the relative costs of multi-storey housing, *Chartered Surveyor* **90** (9), 1958.
126. NISBET, *J., Estimating and Cost Control,* Batsford, 1961.
127. TRICKEY, G. G., Housing cost yardstick technical study, *Architects Journal,* July 16 and 23rd 1975.
128. STONE, P. A., *Building Design Evaluation—Costs-in-Use* (3rd Edition), Spons, 1980.
129. STONE, P. A., The Economics of Building Design, *J. Roy. Statist. Soc.,* Ser. A, **123** (3), 1960.
130. *A Comparative Study of Housing Costs for Different Building Types and Densities,* Building Research Station, Technion, Haifa.
131. STONE, P. A., *The Structure Size and Costs of Urban Settlements,* C.U.P., 1973.
132. STONE, P. A., Design evaluation of a hospital building, *Architects' Journal,* **140** (10).
133. HOLMES, MORVIN, BANKS, KROLL and NEWMAN, *Maintenance costs of flat roofs,* CP4/81, Building Research Establishment, 1981.
134. SCHOFIELD, N. D., Life cycle costs of fire defence in multistorey office buildings, symposium, Quality and Cost in Building, C.I.B., 1980.
135. WOOLARD, F., The economics of multi-storey buildings, *Architectural Science,* Univ. of Sydney, Australia, 1956.
136. GRIMM, C. T. and GROSS, J. C., *Ultimate Costs of Building Walls,* Structural Clay Products Institute, Washington, 1958.
137. JARLE, P., *Rakenteiden yksikkökustannuksia,* Helsinki, 1961.
138. STONE, P. A., Administration and the costs of hospital maintenance, *The Hospital* **60** (7 and 8), 1964.
139. STONE, P. A., Hospital planning and design decisions, *The Hospital* **58** (11), 1962.
140. STONE, P. A., Cost prediction—a guide to design decisions, *Architects Journal,* 2 March 1961.
141. DAVIS, BELFIELD and EVEREST, *Architects' and Builders' Price Book,* Spons, Annual.
142. *Building Maintenance Price Book,* B.M.C.I.S., 1980.
143. CLAPP, M. A., Cost comparisons in Housing Maintenance, *Local Government Finance,* Vol. 67, October, 1963.
144. CULLEN, B. D. and CLAPP, M. A., The maintenance and running costs of school buildings, CP 72/68, Building Research Establishment, 1968.
145. PURKIS, HOW, HOOPER and POOLE, Occupancy costs of offices, Cp 44/77, Building Research Establishment, 1977.
146. CULLEN, B. D. and JEFFERY, I. M., Running costs of hospital buildings, DP 65, Building Research Station, 1967.
147. STONE, P. A., *urban Development in Britain-Costs and Resources,* C.U.P., 1970.
148. CIBULA, E. J., Trends in Timber Supply and Trade, HMSO, 1980.
149. Building Research Station, energy Conservation, C. P., 56/75, Building Research Establishment, Department of the Environment, London, 1975.

150. LEACH, S. J., Energy research and Buildings, BRE News No. 55, Building Research Establishment 1981.

151. BURBERRY, P., ALDRIDGE, L., DAY B., Conserving energy in buildings, *Architects Journal,* Nov. 11th, 1974.

152. Institute of Heating and Ventilating Engineers, I.H.V.E. Guide, London, 1972.

INDEX